AF541082

WASTEWATER TREATMENT ENGINEERING

WASTEWATER TREATMENT ENGINEERING

BY

J.L. BASAK
LOPAMUDRA RAM
GIRISH MAHTO

KNOWLEDGE BAKERS

ISBN: 9789390013432 (HB)

Published by : **Knowledge Bakers**
Pune, India
Email: knowledgebakers2019@gmail.com

Digitally Printed at : **Replika Press Pvt. Ltd.**

PREFACE

Wastewater treatment engineering is a crucial field that deals with the design, construction, and operation of systems and processes used to treat wastewater. The goal of wastewater treatment engineering is to remove pollutants and contaminants from wastewater to make it safe for discharge into the environment or reuse. Wastewater is generated from various sources such as domestic, commercial, and industrial activities. It contains a wide range of pollutants such as organic and inorganic compounds, pathogens, and nutrients. These contaminants can cause significant harm to the environment and human health if discharged without treatment. Wastewater treatment engineering involves various processes, including physical, chemical, and biological treatment processes. Physical processes involve the removal of solids and large particles, while chemical processes use chemicals to remove dissolved pollutants. Biological processes use microorganisms to degrade organic matter and remove nutrients.

Wastewater treatment engineering has become increasingly important in recent years due to the growing global population and the resulting increase in wastewater generation. The need for sustainable and efficient wastewater treatment systems is essential for protecting public health and the environment. The field of wastewater treatment engineering also encompasses the design and operation of wastewater treatment plants, including the selection of appropriate equipment and management of wastewater treatment processes. Wastewater treatment engineering is a critical field that plays a vital role in protecting public health and the environment by designing and operating systems and processes to treat wastewater. It is essential for promoting sustainable development and ensuring a more sustainable and equitable future. Wastewater engineering, also commonly referred to as sanitary engineering, involves the use of mechanical, biological, and chemical operations to achieve and maintain excellent water quality on a daily basis. Wastewater engineers play a pivotal role in our society, for their rigorous testing allows us to always have pure and clean water to drink and use within our homes and businesses. While there's a specific concentration in the transportation and cleaning of irrigation water, blackwater, and greywater, the role of a wastewater engineer is quite diverse.

The present covers a range of topics related to the design, construction, and operation of systems and processes for treating wastewater. It provides an in-depth understanding of various biological and chemical wastewater treatment processes. The book explores the use of bioremediation techniques for the removal of nitroaromatic compounds from wastewater and discusses the selection of bacterial strains for the bioremediation of olive mill wastewater. The book provides perspectives on the biological treatment of sanitary landfill leachate, including the challenges and opportunities associated with this process. The book explores the use of wet air oxidation for the treatment of aqueous wastes and its potential as an alternative to other wastewater treatment processes. The book discusses the modeling of wastewater evolution and the management options available under variable land use scenarios, providing insight into the planning and design of wastewater treatment systems. It is a valuable resource for professionals, researchers, and students in the field of environmental engineering.

We are grateful to all those persons as well as various books, manuals, periodicals, magazines, journals etc. that helped in the preparation of this book. In spite of the best efforts, it is possible that some errors may have occurred into the compilation and editing of the book. Further queries, constructive suggestions and criticisms for the improvement of the book are always welcomed and shall be thankfully acknowledged.

J.L. Basak
Lopamudra Ram
Girish Mahto

CONTENTS

Chapter 1

Biological and Chemical Wastewater Treatment Processes

Abstract

This chapter elucidates the technologies of biological and chemical wastewater treatment processes. The presented biological wastewater treatment processes include: (1) bioremediation of wastewater that includes aerobic treatment (oxidation ponds, aeration lagoons, aerobic bioreactors, activated sludge, percolating or trickling filters, biological filters, rotating biological contactors, biological removal of nutrients) and anaerobic treatment (anaerobic bioreactors, anaerobic lagoons); (2) phytoremediation of wastewater that includes constructed wetlands, rhizofiltration, rhizodegradation, phytodegradation, phytoaccumulation, phytotransformation, and hyperaccumulators; and (3) mycoremediation of wastewater. The discussed chemical wastewater treatment processes include chemical precipitation (coagulation, flocculation), ion exchange, neutralization, adsorption, and disinfection (chlorination/dechlorination, ozone, UV light). Additionally, this chapter elucidates and illustrates the wastewater treatment plants in terms of plant sizing, plant layout, plant design, and plant location.

Keywords: Wastewater treatment, biological treatment, chemical treatment, bioremediation, phytoremediation, mycoremediation, vermifiltration, treatment plant

1. Introduction

The chapter concerns with wastewater treatment engineering, with focus on the biological and chemical treatment processes. It aims at providing a brief and obvious description of the treatment methods, designs, schematics, and specifications. The chapter also answers an important question on how the different processes are interrelated and the correct order of these processes in relation to each other. The main objective of this work was to summarize the work of the eminent scientists in this field in order to provide a clear but concise chapter that can be used as a quick reference for environmental engineers and researchers, and to be effectively implemented in higher education teaching undergraduate and graduate students, as well as extension and outreach.

2. Chapter description and contents overview

The chapter describes the biological and chemical wastewater treatment processes that include:

a. Bioremediation of wastewater using oxidation ponds, aeration lagoons, anaerobic lagoons, aerobic and anaerobic bioreactors, activated sludge, percolating or trickling filters, biological filters, rotating biological contactors, and biological removal of nutrients;

b. Mycoremediation of wastewater using bioreactors;

c. Phytoremediation of wastewater that includes: constructed wetlands, rhizofiltration, rhizodegradation, phytodegradation, phytoaccumulation, phytotransformation, and hyperaccumulators;

d. Vermifiltration and vermicomposting;

e. Microbial fuel cells for electricity production from wastewater;

f. Chemical wastewater treatment processes that include: chemical precipitation, ion exchange, neutralization, adsorption and disinfection (chlorination/dechlorination, ozone, ultraviolet radiation);

g. Wastewater treatment plants. The chapter elucidates and illustrates the plant sizing, plant layout, plant design, and plant location.

3. Overview

3.1. Wastewater treatment techniques

Wastewater, or sewage, originates from human and home wastewaters, industrial wastes, animal wastes, rain runoff, and groundwater infiltration. Generally, wastewater is the flow of used water from a neighborhood. The wastewater consists of 99.9% water by weight, where the remaining 0.1% is suspended or dissolved material. This solid material is a mixture of excrements, detergents, food leftovers, grease, oils, salts, plastics, heavy metals, sands, and grits [1, 2]. Types of wastewaters include: municipal wastewater, industrial wastewaters, mixtures of industrial/domestic wastewaters, and agricultural wastewaters. Typical agricultural industries include: dairy processing industries, meat processing factories, juice and beverage industries, slaughterhouses, vegetable processing facilities, rendering plants, and drainage water of irrigation systems.

Subsequent to primary treatment of wastewater, i.e., physical treatment of wastewater, it still contains large amounts of dissolved and colloidal material that must be removed before discharge. The issue is how to transform the dissolved materials or particulate matters that are too little for sedimentation into larger particles to allow the separation processes to eliminate them. This can be accomplished by secondary treatment, i.e., biological treatment. The treatment of wastewater subsequent to the removal of suspended solids by microorganisms such as algae, fungi, or bacteria under aerobic or anaerobic conditions during which organic matter in wastewater is oxidized or incorporated into cells that can be eliminated by removal process or sedimentation is termed biological treatment. Biological treatment is termed secondary treatment. Chemical treatment, or tertiary treatment, using chemical materials will react with a portion of the undesired chemicals and heavy metals, but a portion of the polluting material will remain unaffected. Additionally, the cost of chemical additives and the environmental problem of disposing large amounts of chemical sludge make this treatment process deficient [1]. Alternatively, the biological treatment must be implemented. This treatment process implements naturally occurring microorganisms to transform the dissolved organic matter into a dense biomass that can be separated from the treated wastewater by the sedimentation process. In fact, the microorganisms utilize the dissolved organic matter as food for themselves, where the generated sludge will be far less for chemical treatment. In practice, therefore, secondary treatment tends to be a biological process with chemical treatment implemented for the removal of toxic compounds.

3.2. Aims of wastewater treatment

The goals of treating the wastewaters are:

a. Transforming the materials available in the wastewater into secure end products that are able to be safely disposed off into domestic water devoid of any negative environmental effects;

b. Protecting public health;

c. Ensuring that wastewaters are efficiently handled on a trustworthy basis without annoyance or offense;

d. Recycling and recovering the valuable components available in wastewaters;

e. Affording feasible treatment processes and disposal techniques;

f. Complying with the legislations, acts and legal standards, and approval conditions of discharge and disposal.

3.3. Biological treatment processes

The secondary treatment can be defined as "treatment of wastewater by a process involving biological treatment with a secondary sedimentation". In other words, the secondary treatment is a biological process. The settled wastewater is introduced into a specially designed bioreactor where under aerobic or anaerobic conditions the organic matter is utilized by microorganisms such as bacteria (aerobically or anaerobically), algae, and fungi (aerobically). The bioreactor affords appropriate bioenvironmental conditions for the microorganisms to reproduce and use the dissolved organic matter as energy for themselves. Provided that oxygen and food, in the form of settled wastewater, are supplied to the microorganisms, the biological oxidation process of dissolved organic matter will be maintained. The biological process is mostly carried out bacteria that form the basic trophic level (the level of an organism is the position it occupies in a food chain) of the food chain inside the bioreactor. The bioconversion of dissolved organic matter into thick bacterial biomass can fundamentally purify the wastewater. Subsequently, it is crucial to separate the microbial biomass from the treated wastewater though sedimentation. This secondary sedimentation is basically similar to primary sedimentation except that the sludge contains bacterial cells rather than fecal solids. The biological removal of organic matter from settled wastewater is conducted by microorganisms, mainly heterotrophic bacteria but also occasionally fungi. The microorganisms are able to decompose the organic matter through two different biological processes: biological oxidation and biosynthesis [1]. The biological oxidation forms some end-products, such as minerals, that remain in the solution and are discharged with the effluent (Eq. 1). The biosynthesis transforms the colloidal and dissolved organic matter into new cells that form in turn the dense biomass that can be then removed by sedimentation (Eq. 2). Figure 1 summarizes these processes. On the other hand, algal photosynthesis plays an important role in some cases (Figure 2).

Oxidation:

$$\underset{\text{(Organic matter)}}{COHNS} + O_2 + \text{Bacteria} \rightarrow CO_2 + \underset{\text{+Energy}}{NH_3} + \text{Other end products} \quad (1)$$

Biosynthesis:

$$\underset{\text{(Organic matter)}}{COHNS + O_2} + \text{Bacteria} \rightarrow \underset{\text{(New cells)}}{C_5H_7NO_2} \quad (2)$$

3.3.1. Useful terms

The following terms are the most used in biological treatment processes [2]:

a. DO: Dissolved Oxygen (mg L^{-1})

b. BOD: Biochemical Oxygen Demand (mg L^{-1})

c. BOD_5: BOD (mg L^{-1}), incubation at 15°C for 5 days

d. COD: Chemical Oxygen Demand (mg L^{-1})

e. CBOD: Carbonaceous BOD (mg L^{-1})

f. NBOD: Nitrogenous (mg L^{-1})

g. SOD: Sediment Oxygen Demand (mg L^{-1})

h. TBOD: Total BOD (mg L^{-1})

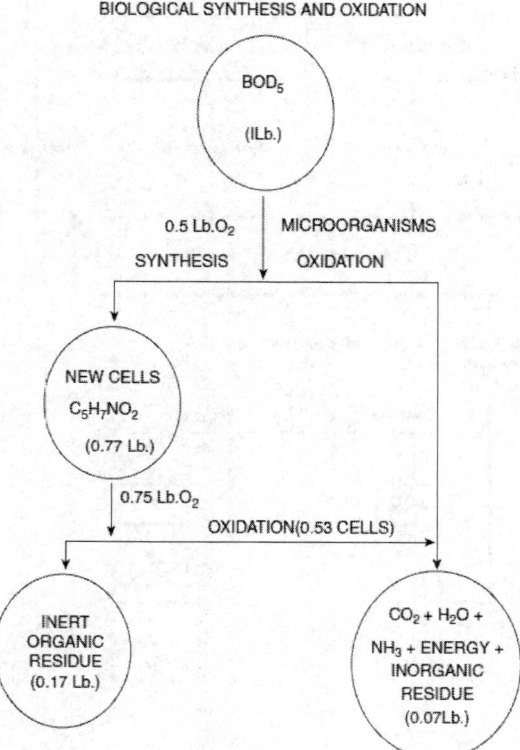

Figure 1. Biological synthesis and oxidation [3].

3.4. Chemical treatment processes

In early wastewater treatment technologies, chemical treatment has preceded biological treatment. Recently, the biological treatment precedes chemical treatment in the treatment process. Chemical treatment is now considered as a tertiary treatment that can be more broadly defined as "treatment of wastewater by a process involving chemical treatment". The mostly implemented chemical treatment processes are: chemical precipitation, neutralization, adsorption, disinfection (chlorine, ozone, ultraviolet light), and ion exchange.

4. Biological treatment of wastewater

4.1. Biological growth equation

The biological growth can be described according to the Monod equation:

$$\mu = (\lambda S) / (K_S + S)$$

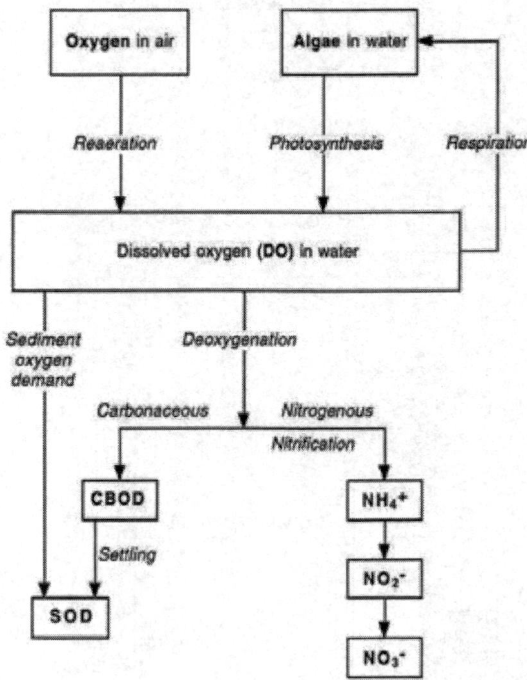

Figure 2. Photosynthesis and oxidation [2].

Where, μ is the specific growth rate coefficient; λ is the maximum growth rate coefficient that occurs at 0.5 μ_{max}; S is the concentration of limiting nutrient, that is BOD and COD; and K_S is the Monod coefficient [3].

Generally, the bacterial growth can be explained by the following simplified figure:

$$\text{Organics+Bacteria+Nutrients+Oxygen} \\ \rightarrow \text{New Bacteria+}CO_2\text{+}H_2O\text{+Residual Organics+Inorganics}$$

Several bioenvironmental factors affect the activity of bacteria and the rate of biochemical reactions. The most important factors are: temperature, pH, dissolved oxygen, nutrient concentration, and toxic materials. All these factors can be controlled within a biological treatment system and/or a bioreactor in order to ensure that the microbial growth is maintained under optimum bioenvironmental conditions. The majority of biological treatment systems operate in the mesophilic temperature range, where the optimal temperature ranges from 20°C to 40°C. Aeration tanks and percolating filters operate at the temperature of the wastewater that ranges from 12°C to 25°C; although in percolating filters, the air temperature and the ventilation rate may have a significant effect on heat loss. The higher temperatures increase the biological activity and metabolism, which result in increasing the substrate removal rate.

However, the increased metabolism at the higher temperatures may lead to problems of oxygen limitations.

4.2. Bacterial kinetics

The bacterial kinetics can be shown in Figures 3 and 4. The microbial growth curve that shows bacterial density and specific growth rate at the different growth phases is shown in Figure 3. The microbial growth curves that compare the total biomass and the variable biomass are shown in Figure 4.

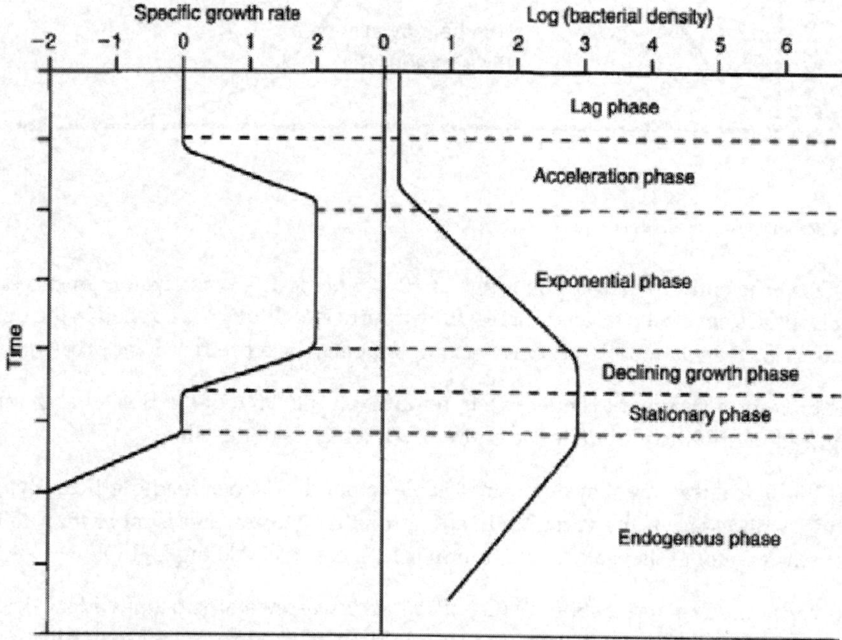

Figure 3. Microbial growth curve [1].

4.3. Principles of biological treatment

The principles of biological treatment of wastewater were stated by [3]. The following is a summary of the principles:

1. The biological systems are very sensitive for extreme variations in hydraulic loads. Diurnal variations of greater than 250% are problematic because they will create biomass loss in the clarifiers.

2. The growth rate of microorganisms is highly dependent on temperature. A 10°C reduction in wastewater temperature dramatically decreases the biological reaction rates to half.

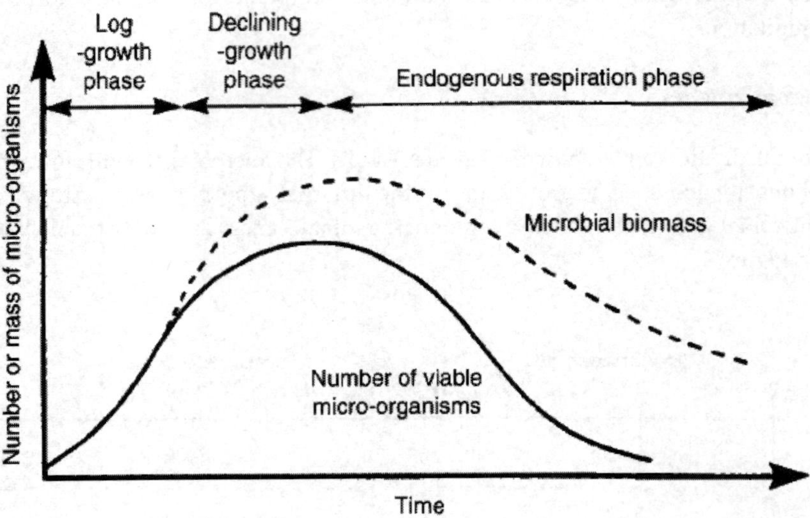

Figure 4. Microbial growth curves [1].

3. BOD is efficiently treated in the range of 60 to 500 mg L^{-1}. Wastewaters in excess of 500 mg L^{-1} BODs have been treated successfully if sufficient dilution is applied in the treatment process, or if an anaerobic process was implemented as a pretreatment process.

4. The biological treatment is effective in removing up to 95% of the BOD. Large tanks are required in order to eliminate the entire BOD, which is not feasible.

5. The biological treatment systems are unable to handle "shock loads" efficiently. Equalization is necessary if the variation in strength of the wastewater is more than 150% or if that wastewater at its peak concentration is in excess of 1,000 mg L^{-1} BOD.

6. The carbon:nitrogen:phosphorus (C:N:P) ratio of wastewater is usually ideal. The C:N:P ratio of industrial wastewaters should range from 100:20:1 to 100:5:1 for a most advantageous biological process.

7. If the C:N:P ratio of the wastewater is strong in an element in comparison to the other elements, then poor treatment will result. This is especially true if the wastewater is very strong in carbon. The wastewater should also be neither very weak nor very strong in an element; although very weak is acceptable, it is difficult to treat.

8. Oils and solids cannot be handled in a biological treatment system because they negatively affect the treatment process. These wastes should be pretreated to remove solids and oils.

9. Toxic and biological-resistant materials require special consideration and may require pretreatment before being introduced into a biological treatment system.

10. Although the capacity of the wastewater to utilize oxygen is unlimited, the capacity of any aeration system is limited in terms of oxygen transfer.

4.4. Bioremediation of wastewater

Bioremediation is a treatment process that involves the implementation of microorganisms to remove pollutants from a contaminated setting. Bioremediation can be defined as "treatment that implements natural organisms to decompose hazardous materials into less toxic or nontoxic materials". Some examples of bioremediation-related technologies are phytoremediation, bioaugmentation, rhizofiltration, and biostimulation. The microorganisms implemented to carry out the bioremediation are called bioremediators. However, some pollutants are not easily removed or decomposed by bioremediation. For example, heavy metals such as lead and cadmium are not eagerly captured by bioremediators. Example of bioremediation: fish bone char has been shown to bioremediate small amounts of cadmium, copper, and zinc.

The bioremediation of wastewater can be achieved by autotrophs or heterotrophs. A heterotroph is an organism that is unable to fix carbon and utilizes organic carbon for its growth. Heterotrophs are divided based on their source of energy. If the heterotroph utilizes light as its source of energy, then it is considered a photoheterotroph. If the heterotroph utilizes organic and/or inorganic compounds as energy sources, it is then considered a chemoheterotroph. Autotrophs, such as plants and algae, that are able to utilize energy from sunlight are called photoautotrophs. Autotrophs that utilize inorganic compounds to produce organic compounds such as carbohydrates, fats, and proteins from inorganic carbon dioxide are called lithoautotrophs. These reduced carbon compounds can be utilized as energy sources by autotrophs and provide the energy in food consumed by heterotrophs. Over 95% of all organisms are heterotrophic.

4.4.1. Aerobic treatment

Aeration has been used to remove trace organic volatile compounds (VOCs) in water. It has also been employed to transfer a substance, such as oxygen, from air or a gas phase into water in a process called "gas adsorption" or "oxidation", i.e., to oxidize iron and/or manganese. Aeration also provides the escape of dissolved gases, such as CO_2 and H_2S. Air stripping has been also utilized effectively to remove NH_3 from wastewater and to remove volatile tastes and other such substances in water [2]. Samer [4] and Samer et al. [5] mentioned that aerobic treatment with biowastes is effective in reducing harmful gaseous emissions as greenhouse gases (CH_4 and N_2O) and ammonia.

4.4.1.1. Oxidation ponds

Oxidation ponds (Figure 5) are aerobic systems where the oxygen required by the heterotrophic bacteria (a heterotroph is an organism that cannot fix carbon and uses organic carbon for growth) is provided not only by transfer from the atmosphere but also by photosynthetic algae. The algae are restricted to the euphotic zone (sunlight zone), which is often only a few centimeters deep. Ponds are constructed to a depth of between 1.2 and 1.8 m to ensure maximum penetration of sunlight, and appear dark green in color due to dense algal development. Samer [6] and Samer et al. [7] illustrated the structures and constructions of the aerobic treatment tanks and the used building materials.

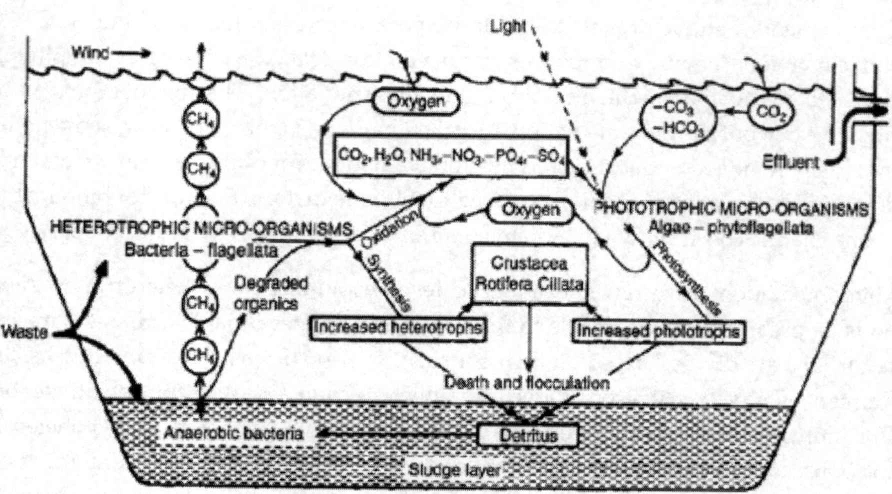

Figure 5. Aerobic system/oxidation pond [1].

In oxidation ponds, the algae use the inorganic compounds (N, P, CO_2) released by aerobic bacteria for growth using sunlight for energy. They release oxygen into the solution that in turn is utilized by the bacteria, completing the symbiotic cycle. There are two distinct zones in facultative ponds: the upper aerobic zone where bacterial (facultative) activity occurs and a lower anaerobic zone where solids settle out of suspension to form a sludge that is degraded anaerobically.

4.4.1.2. Aeration lagoons

Aeration lagoons are profound (3–4 m) compared to oxidation ponds, where oxygen is provided by aerators but not by the photosynthetic activity of algae as in the oxidation ponds. The aerators keep the microbial biomass suspended and provide sufficient dissolved oxygen that allows maximal aerobic activity. On the other hand, bubble aeration is commonly used where the bubbles are generated by compressed air pumped through plastic tubing laid through the base of the lagoon. A predominately bacterial biomass develops and, whereas there is neither sedimentation nor sludge return, this procedure counts on adequate mixed liquor formed in the tank/lagoon. Therefore, the aeration lagoons are suitable for strong but degradable wastewater such as wastewaters of food industries. The hydraulic retention time (HRT) ranges from 3 to 8 days based on treatment level, strength, and temperature of the influent. Generally, HRT of about 5 days at 20°C achieves 85% removal of BOD in household wastewater. However, if the temperature falls by 10°C, then the BOD removal will decrease to 65% [1].

4.4.2. *Anaerobic treatment*

The anaerobic treatments are implemented to treat wastewaters rich in biodegradable organic matter (BOD >500 mg L^{-1}) and for further treatment of sedimentation sludges. Strong organic wastewaters containing large amounts of biodegradable materials are discharged mainly by agricultural and food processing industries. These wastewaters are difficult to be treated aerobically due to the troubles and expenses of fulfillment of the elevated oxygen demand to preserve the aerobic conditions [1]. In contrast, anaerobic degradation occurs in the absence of oxygen. Although the anaerobic treatment is time-consuming, it has a multitude of advantages in treating strong organic wastewaters. These advantages include elevated levels of purification, aptitude to handle high organic loads, generating small amounts of sludges that are usually very stable, and production of methane (inert combustible gas) as end-product.

Anaerobic digestion is a complex multistep process in terms of chemistry and microbiology. Organic materials are degraded into basic constituents, finally to methane gas under the absence of an electron acceptor such as oxygen [8]. The basic metabolic pathway of anaerobic digestion is shown in Figures 6 and 7. To achieve this pathway, the presence of very different and closely dependent microbial population is required.

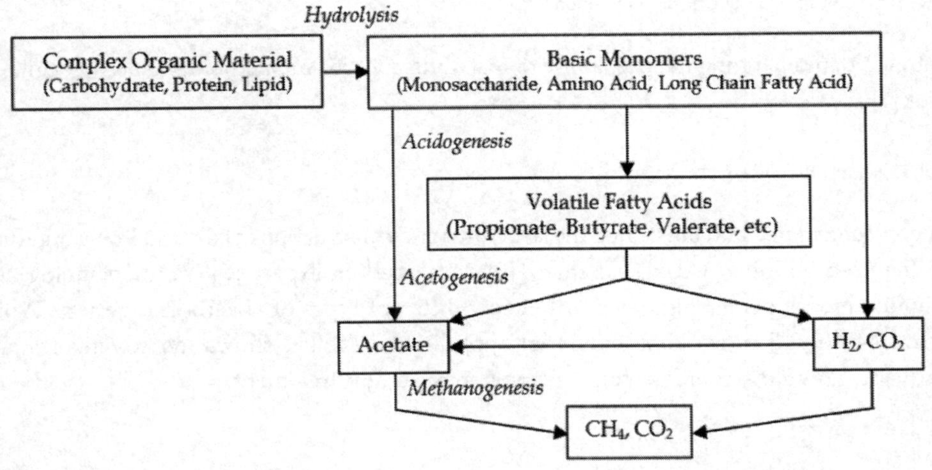

Figure 6. Steps of the anaerobic digestion process [8].

Suitable wastewaters include livestock manure, food processing effluents, petroleum wastes (if the toxicity is controlled), and canning and dyestuff wastes where soluble organic matters are implemented in the treatment. Most anaerobic processes (solids fermentation) occur in two predetermined temperature ranges: mesophilic or thermophilic. The temperature ranges are 30–38°C and 38–50°C, respectively [3]. In contrast to aerobic systems, absolute stabilization of organic matter is not achievable under anaerobic conditions. Therefore, subsequent aerobic treatment of the anaerobic effluents is usually essential. The final waste matter discharged by the anaerobic treatment includes solubilized organic matter that is acquiescent to aerobic

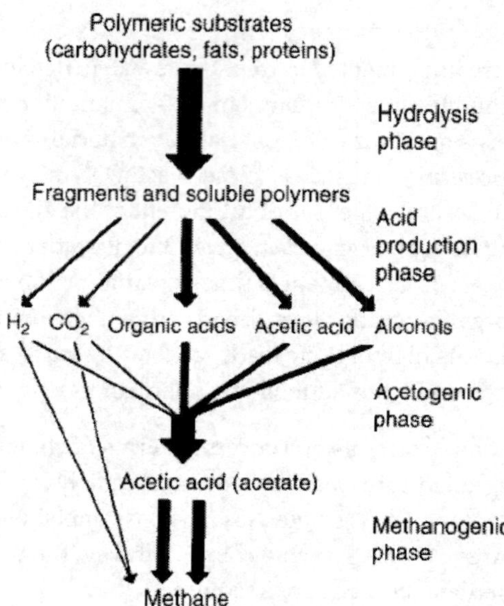

Figure 7. Major steps in anaerobic decomposition [1].

treatment demonstrating the possibility of installing collective anaerobic and aerobic units in series [1].

4.4.2.1. Anaerobic digesters

Samer [9] elucidated and illustrated the structures and constructions of the anaerobic digesters and the used building materials. Samer [10] developed an expert system for planning and designing biogas plants. Figures 8 to 13 show different types of anaerobic digesters. While Figures 14 and 15 show some industrial applications. Table 1 shows the advantages and disadvantages of anaerobic treatment compared to aerobic treatment.

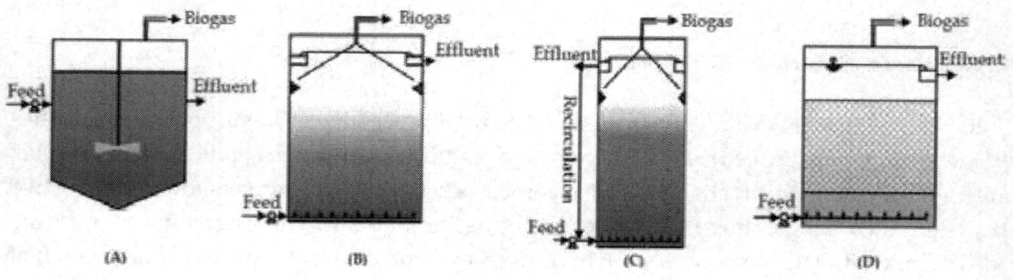

Figure 8. Most commonly used anaerobic reactor types: (A) Completely mixed anaerobic digester, (B) UASB reactor, (C) AFB or EGSB reactor, and (D) Upflow AF [8].

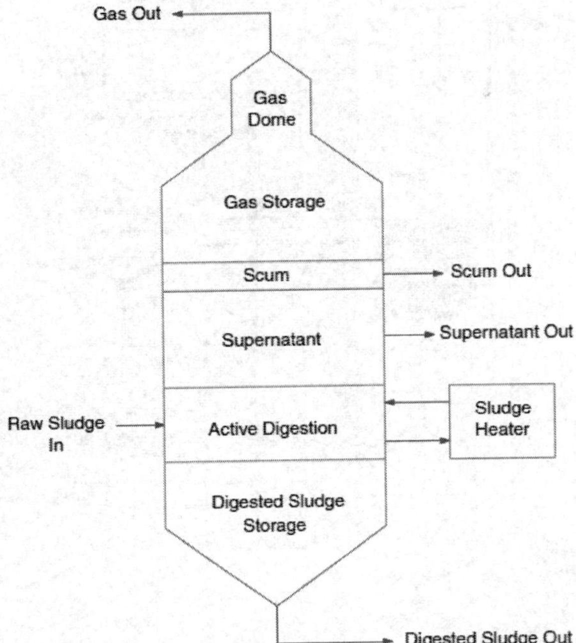

Figure 9. Single-stage conventional anaerobic digester [3].

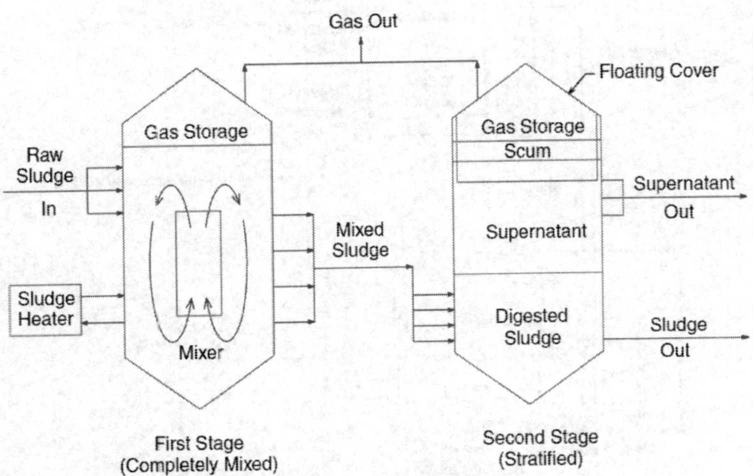

Figure 10. Dual-stage high rate digester [3].

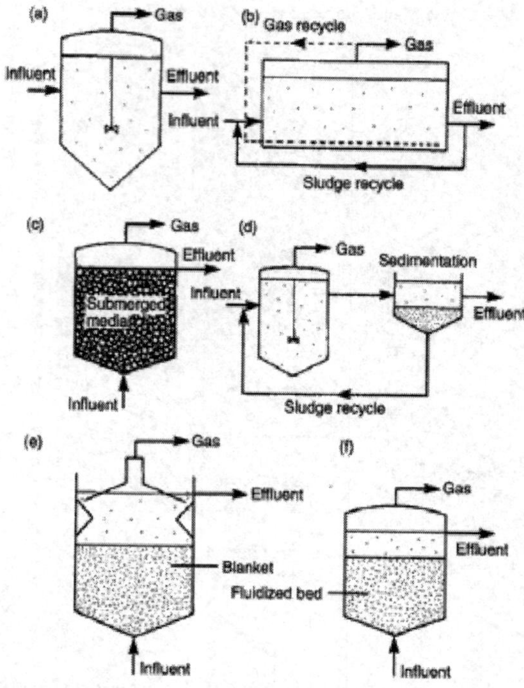

Figure 11. Schematic representation of digester types. Flow-through (A–B) and contact systems (C–F) [1].

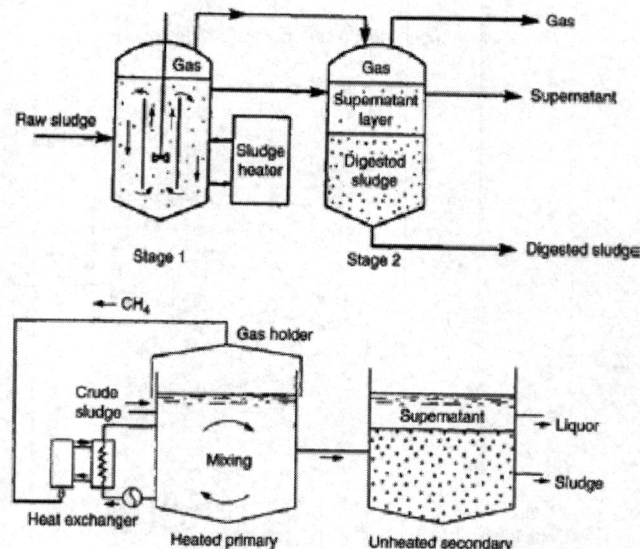

Figure 12. The upper scheme shows a two-stage anaerobic sludge digester, while the lower scheme shows the conventional sludge digestion plant [1].

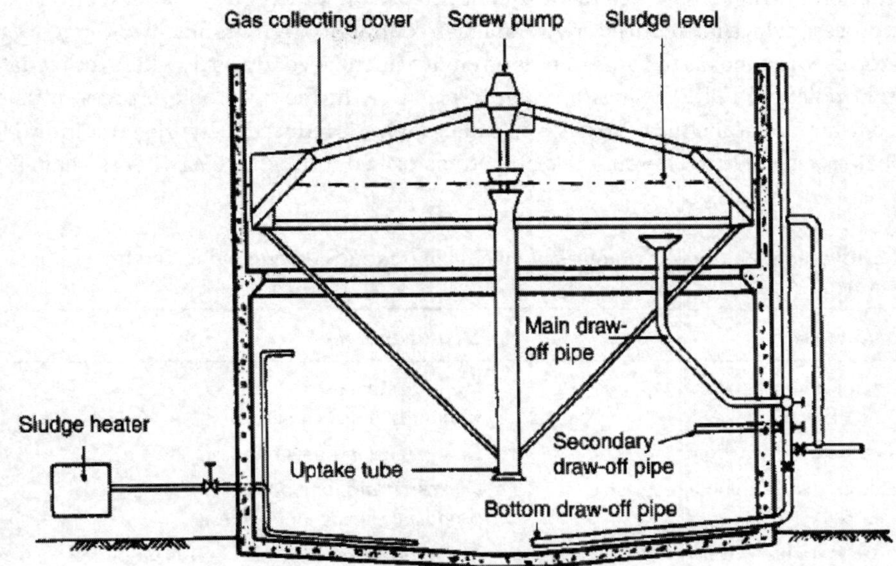

Figure 13. Primary digestion tank with screw mixing pump and external heater [1].

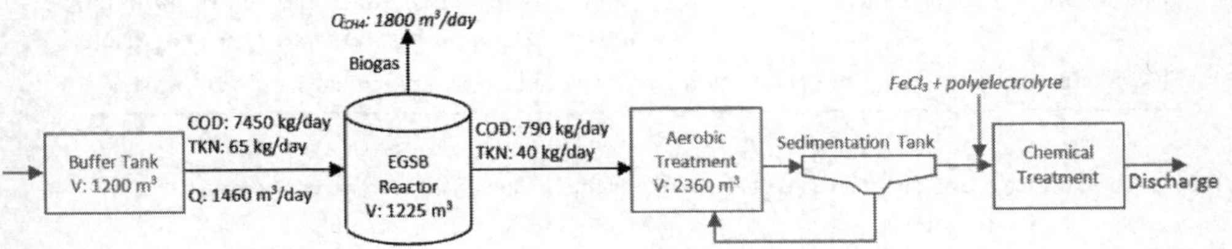

Figure 14. Wastewater treatment plant for corn processing industry [8].

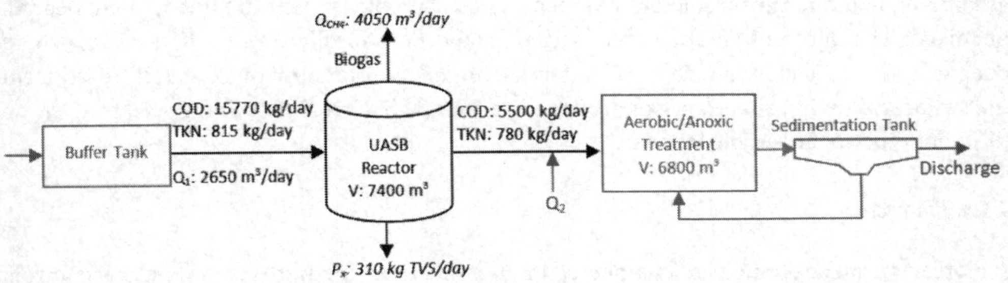

Figure 15. Mass balance study for a wastewater treatment plant of the baker's yeast industry [8].

By definition, the anaerobic treatment is conducted without oxygen. It is different from an anoxic process, which is a reduced environment in contrast to an environment without oxygen. Both processes are anoxic, but anaerobic is an environment beyond anoxic where the oxidation reduction potential (ORP) values are highly negative. In the anaerobic process, nitrate is reduced to ammonia and nitrogen gas, and sulfate (SO_3^{2-}) is reduced to hydrogen sulfide (H_2S). Phosphate is also reduced because it is often transformed through the ADP–ATP chain [3].

The advantages and disadvantages of anaerobic treatment compared to aerobic treatment

Advantages	*Disadvantages*
Low operational costs	High capital costs Generally require heating
Low sludge production	Low retention times required (>24 h)
Reactors sealed giving no odour or aerosols	Corrosive and malodorous compounds produced during anaerobiosis
Sludge is highly stabilized	Not as effective as aerobic stabilization for pathogen destruction
Methane gas produced as end product	Hydrogen sulphide also produced
Low nutrient requirement due to lower growth rate of anaerobes	Reactor may require additional alkalinity
Can be operated seasonally	Slow growth rate of anaerobes can result in long initial start-up of reactors and recovery periods
Rapid start-up possible after acclimation	Only used as pre-treatment for liquid wastes

Table 1. The advantages and disadvantages of anaerobic treatment compared to aerobic treatment [1].

4.4.2.2. Anaerobic lagoons

An anaerobic lagoon is a deep lagoon, fundamentally without dissolved oxygen, that enforces anaerobic conditions. The anaerobic process occurs in deep ground ponds, and such basins are implemented for anaerobic pretreatment. The anaerobic lagoons are not aerated, heated, or mixed. The depth of an anaerobic lagoon should be typically deeper than 2.5 m, where deeper lagoons are more efficient. Such depths diminish the amount of oxygen diffused from the surface, allowing anaerobic conditions to prevail (U.S. EPA, 2002). Figures 16 to 18 show different types of anaerobic lagoons.

4.4.3. Bioreactors

A bioreactor can be defined as "engineered or manufactured apparatus or system that controls the embraced or encompassed bioenvironment". Precisely, the bioreactor is a vessel in which a biochemical process is conducted, where it involves microorganisms (e.g., bacteria, algae, fungi) or biochemical substances (e.g., enzymes) derived from such microorganisms. The

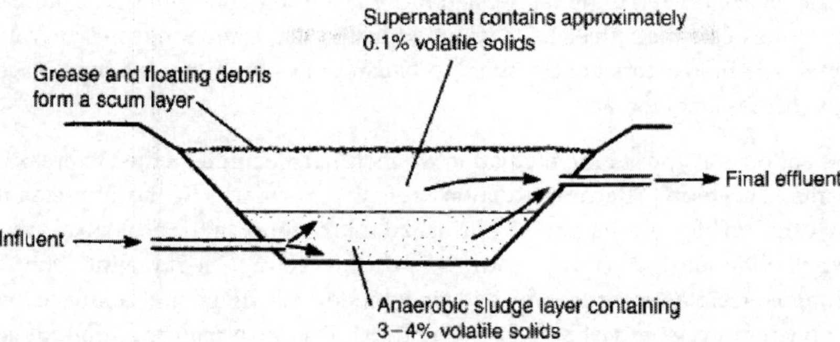

Figure 16. Anaerobic lagoon for strong wastewater treatment, such as meat processing wastewater [1].

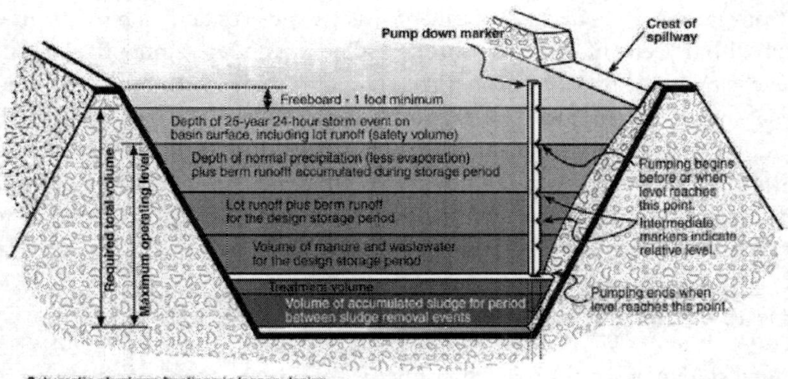

Figure 17. Schematic of volume fractions in anaerobic lagoon design [11].

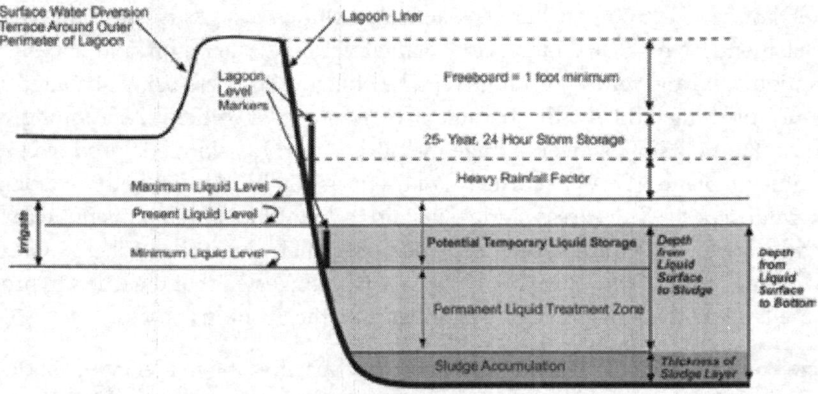

Figure 18. Anaerobic wastewater treatment lagoon [12].

treatment can be conducted under either aerobic or anaerobic conditions. The bioreactors are commonly made of stainless steel, usually cylindrical in shape and range in size from liters to cubic meters. The bioreactors are classified as batch, plug, or continuous flow reactors (e.g., continuous stirred-tank bioreactor).

Mycoremediation is a type of bioremediation where fungi are implemented to break down the contaminants. The term "mycoremediation" refers particularly to the implementation of fungal "mycelia" in bioremediation. The principal role of fungi in the ecological system is the breakdown of pollutants, which is performed by the mycelium. The mycelium, the vegetative part of a fungus, secretes enzymes and acids that biodegrade lignin and cellulose that are the main components of vegetative fibers. Lignin and cellulose are organic compounds composed of long chains of carbon and hydrogen, and therefore they are structurally similar to several organic pollutants. One key issue is specifying the right fungus to break down a determined pollutant. Similarly, mycofiltration is a process that uses fungal mycelia to filter toxic compounds from wastewater. In an experiment, wastewater contaminated with diesel oil was inoculated with mycelia of oyster mushrooms. One month later, more than 93% of many of the polycyclic aromatic hydrocarbons (PAH) had been reduced to non-toxic components in the mycelial-inoculated samples. The natural microbial community participates with the fungi to break down contaminants, eventually into CO_2 and H_2O. Wood-degrading fungi are particularly effective in breaking down aromatic pollutants (toxic components of petroleum), as well as chlorinated compounds (certain persistent pesticides). Figures 19 to 22 show different types and designs of bioreactors.

4.4.4. Activated sludge

The activated sludge process is based on a mixture of thick bacterial population suspended in the wastewater under aerobic conditions. With unlimited nutrients and oxygen, high rates of bacterial growth and respiration can be attained, which results in the consumption of the available organic matter to either oxidized end-products (e.g., CO_2, NO_3^-, SO_4^{2-}, and PO_4^{3-}) or biosynthesis of new microorganisms. The activated sludge process is based on five interdependent elements, which are: bioreactor, activated sludge, aeration and mixing system, sedimentation tank, and returned sludge [1]. The biological process using activated sludge is a commonly used method for the treatment of wastewater, where the running costs are inexpensive (Figure 23). However, a huge quantity of surplus sludge is produced in wastewater treatment plants (WWTPs) which is an enormous burden in both economical and environmental aspects. The excess sludge contains a lot of moisture and is not easy to treat. The byproducts of WWTPs are dewatered, dried, and finally burnt into ashes. Some are used in farm lands as compost fertilizer [15]. However, it is suggested that the dried byproducts of WWTPs are fed into the pyrolysis process rather than the burning process.

The sludge volume index (SVI) is an estimation that specifies the tendency of aerated solids, i.e., activated sludge solids, to become dense or concentrated through the thickening process. SVI can be computed as follows: (a) allowing a mixed liquor sample from the aeration tank to sediment in 30 min; (b) determining the concentration of the suspended solids for a sample of the same mixed liquor; (c) SVI is then computed as ratio of the measured wet volume (mL/L)

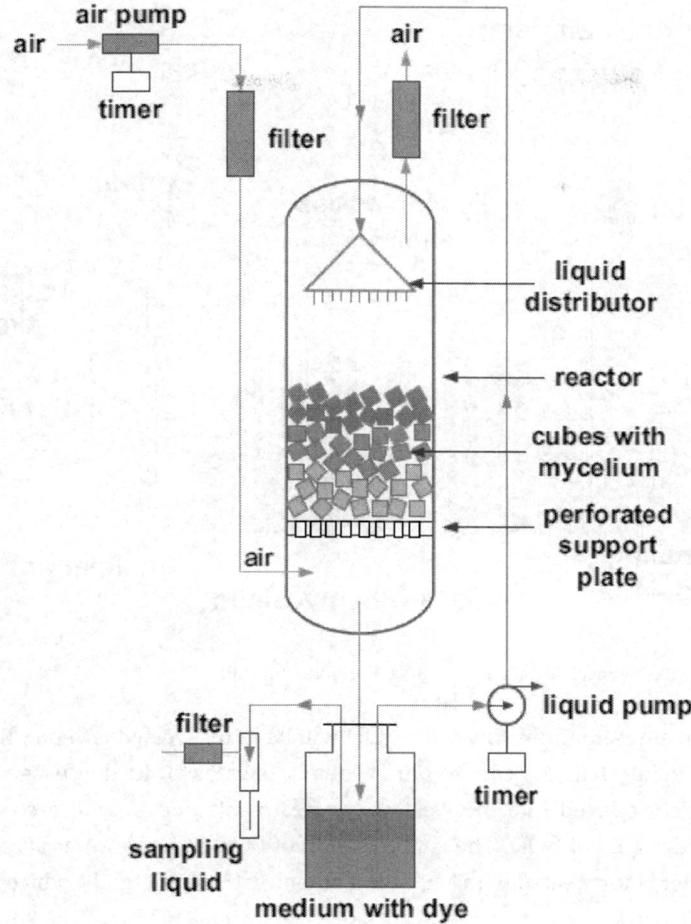

Figure 19. A bioreactor for fungal degradation: trickle bed bioreactor [13].

of the settled sludge to the dry weight concentration of MLSS in g/L (Source: Office of Water Programs, Sacramento State, USA).

During the treatment of wastewater in aeration tanks through the activated sludge process (Table 2) there are suspended solids, where the concentration of the suspended solids is termed as mixed liquor suspended solids (MLSS), which is measured in milligrams per liter (mg L^{-1}). Mixed liquor is a mixture of raw wastewater and activated sludge in an aeration tank. MLSS consists mainly of microorganisms and non-biodegradable suspended solids. MLSS is the effective and active portion of the activated sludge process that ensures that there is adequate quantity of viable biomass available to degrade the supplied quantity of organic pollutants at any time. This is termed as Food to Microorganism Ratio (F/M Ratio) or food to mass ratio. If this ratio is kept at the suitable level, then the biomass will be able to consume high quantities of the food, which reduces the loss of residual food in the discharge. In other words, the more

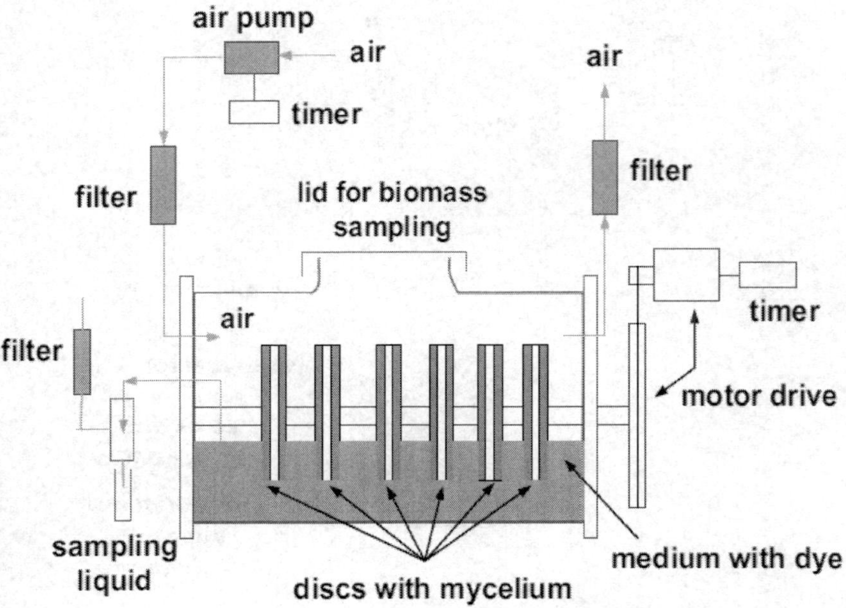

Figure 20. A bioreactor for fungal degradation: rotating disc bioreactor [13].

the biomass consumes food the lower the BOD will be in the treated effluent. It is important that MLSS eliminates BOD in order to purify the wastewater for further usage and hygiene. Raw sewage is introduced into the wastewater treatment process with a concentration of several hundred mg L^{-1} of BOD. The concentration of BOD in wastewater is reduced to less than 2 mg L^{-1} after being treated with MLSS and other treatment methods, which is considered to be safe water to use.

Specification	Value	Unit
BOD-Sludge Loading	0.40	mg L^{-1}
BOD-Volume Loading	0.20	mg L^{-1}
MLSS	2000	mg L^{-1}
COD of Influent	300	mg L^{-1}
Amount of Influent	4.48	L d^{-1}
Aeration Rate	3.00	L min^{-1}

Table 2. Conventional activated sludge [15].

The biological treatment process is the most commonly implemented method for the treatment of domestic sewage. This method implements bacterial populations that possess superior

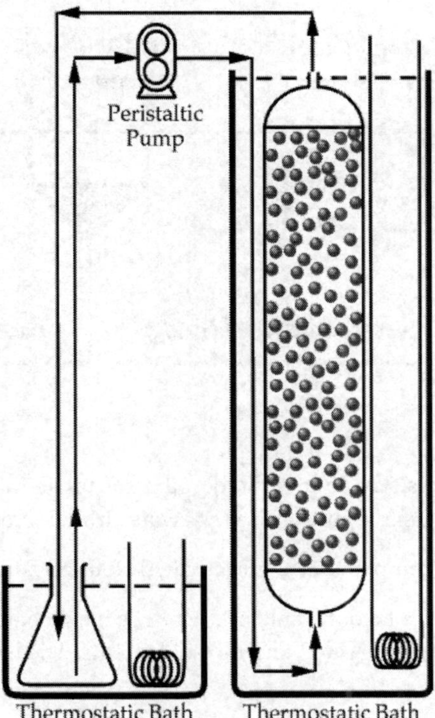

Figure 21. Fluidized bed bioreactor [14].

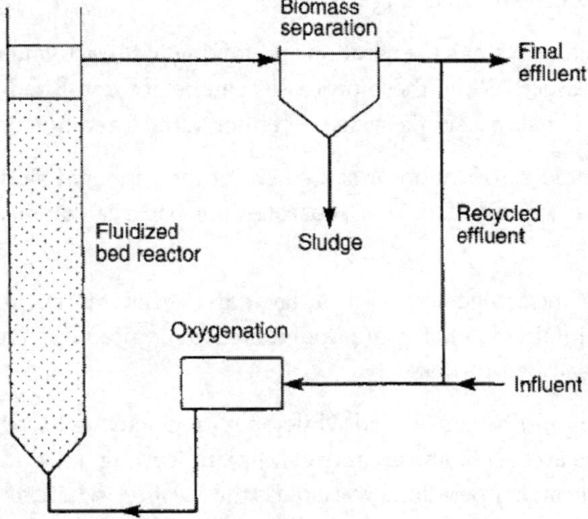

Figure 22. Typical design of fluidized bed reactor system [1].

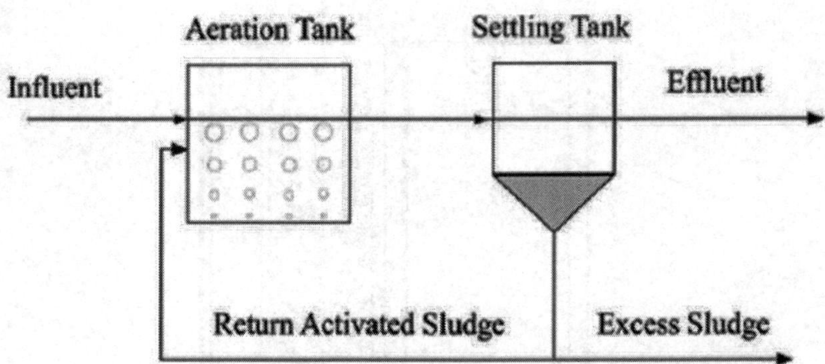

Figure 23. Activated sludge [15].

sedimentation characteristics. The living microorganisms break down the organic matter in the wastewater and consequently purify the wastewater from biological waste [15].

According to [1], the main components of all activated sludge systems are:

1. The bioreactor: it can be a lagoon, tank, or ditch. The main characteristic of a bioreactor is that it contains sufficiently aerated and mixed contents. The bioreactor is also known as the aeration tank.

2. Activated sludge: it is the bacterial biomass inside the bioreactor that consists mostly of bacteria and other flora and microfauna. The sludge is a flocculent suspension of these microorganisms and is usually termed as the mixed liquor suspended solids (MLSS) that ranges between 2,000 and 5,000 mg L^{-1}.

3. Aeration and mixing system: the aeration and mixing of the activated sludge and the raw influent are necessary. While these processes can be accomplished separately, they are usually conducted using a single system of either surface aeration or diffused air.

4. Sedimentation tank: clarification or settlement of the activated sludge discharged from the aeration tank is essential. This separates the bacterial biomass from the treated wastewater.

5. Returned sludge: the settled activated sludge in the sedimentation tank is returned to the bioreactor to maintain the microbial population at a required concentration to guarantee persistence of treatment process.

Several parameters should be considered while operating activated sludge plants. The most important parameters are: (1) biomass control, (2) plant loading, (3) sludge settleability, and (4) sludge activity. The main operational variable is the aeration, where its major functions are: (1) ensuring a sufficient and continuous supply of dissolved oxygen (DO) for the bacterial population, (2) keeping the bacteria and the biomass suspended, and (3) mixing the influent wastewater with the biomass and removing from the solution the excessive CO_2 resulting from oxidation of organic matter [1].

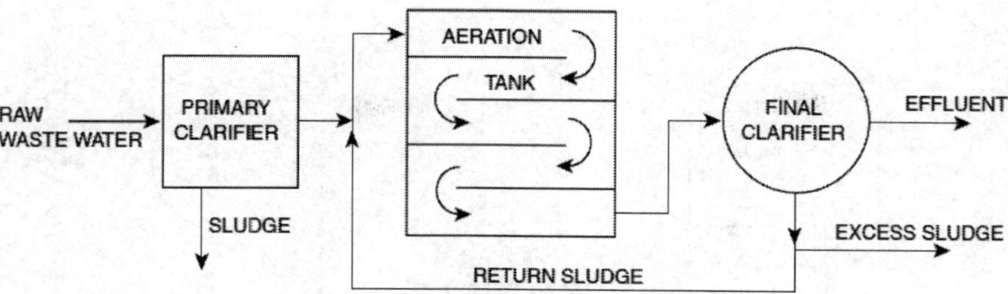

Figure 24. Conventional activated sludge plant [3].

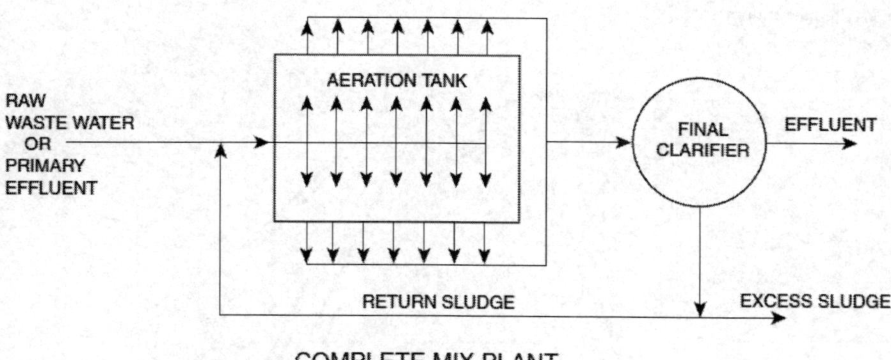

Figure 25. Complete mix plant [3].

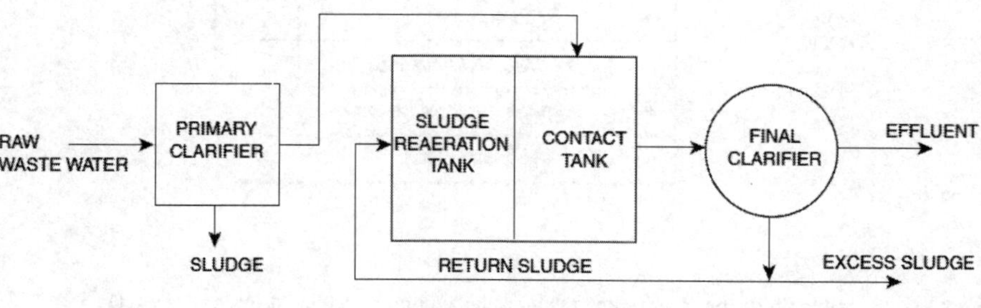

Figure 26. Contact stabilization plant [3].

There are several types of activated sludge processes, e.g., conventional activated sludge plant (Figure 24), complete mix plant (Figure 25), contact stabilization plant (Figure 26), and step aeration plant (Figure 27). Figure 28 shows the food pyramid that represents the feeding relationships within the activated sludge process.

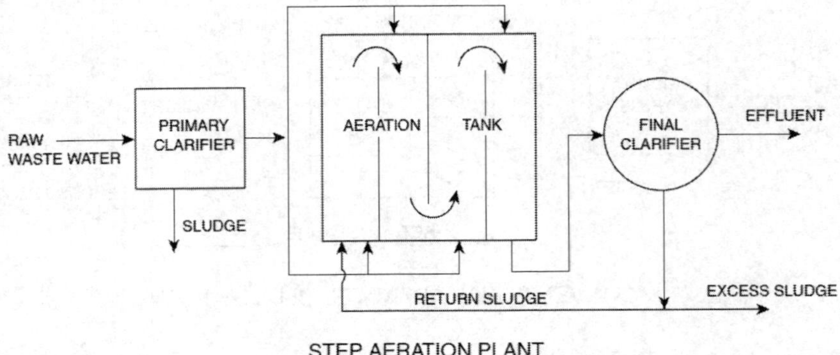

Figure 27. Step aeration plant [3].

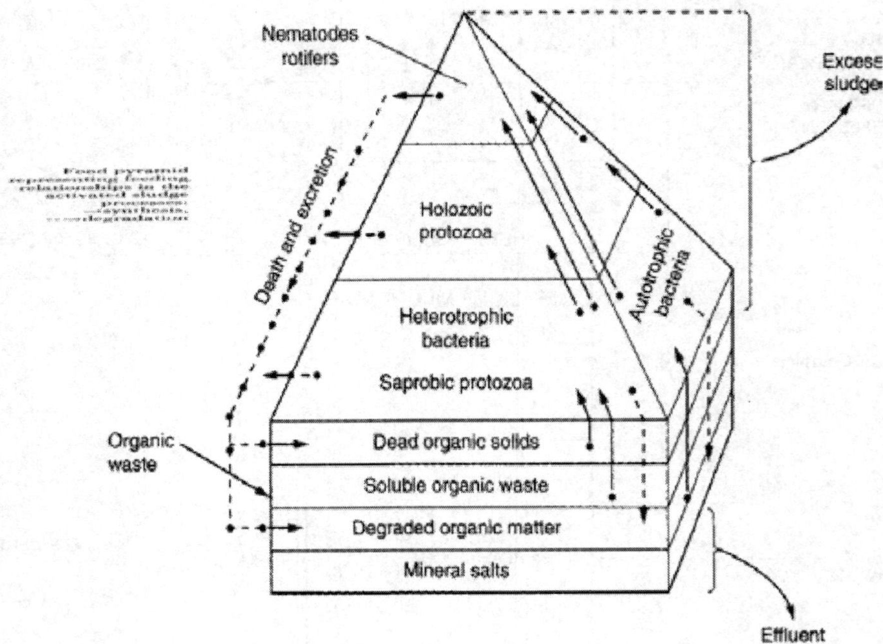

Figure 28. Food pyramid illustrating the feeding relationships within the activated sludge process [1].

4.4.5. Biological filters

The main systems of operation of biological filters are: (a) single filtration, (b) recirculation, (c) ADF, and (d) two-stage filtration with high-rate primary biotower (Figure 29). There are several types of biological filters, for example, submerged aerated filters that are widely known as biological aerated filters (BAFs) and are the commonly implemented design (Figure 30), and the percolating (trickling) filters (Figure 31). The BAFs implement either the sunken granular media with upward (Figure 30a) or download (Figure 30b) flows, or floating granular media with upward flow (Figure 30c), which is the most common design of BAFs. In order to compare the biological filters and the activated sludge systems (Figures 31 and 32), the comparison is based on the oxidation that can be accomplished by three processes:

1. Spreading the wastewater into a thin film of liquid with a large surface area, consequently the required oxygen can be supplied by gaseous diffusion, which is the case of the percolating filters.
2. Aerating the wastewater by pumping air in the form of bubbles or stirring forcefully, which is the case of the activated sludge process.
3. Implementing algae to produce oxygen by photosynthesis, which is the case of the stabilization ponds.

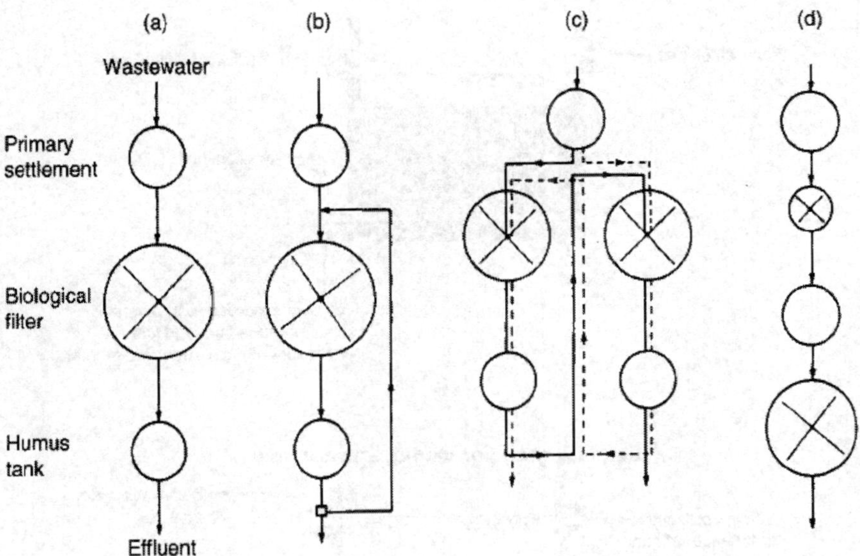

Figure 29. The main systems of operation of biological filters [1].

4.4.6. Rotating biological contactors

The rotating biological contactors (RBC) system (Figure 33) can be implemented to amend and improve the available treatment processes as the secondary or tertiary treatment processes. The RBC is successfully implemented in all three steps of the biological treatment, which are BOD_5 removal, nitrification, and denitrification. The process is a fixed-biofilm of either aerobic or anaerobic biological treatment system for removal of nitrogenous and carbonaceous compounds from wastewater (Figure 34). The RBC installations (Figure 35) were designed for removal of BOD_5 or ammonia nitrogen (NH_3-N), or both, from wastewater [1, 2].

(a)

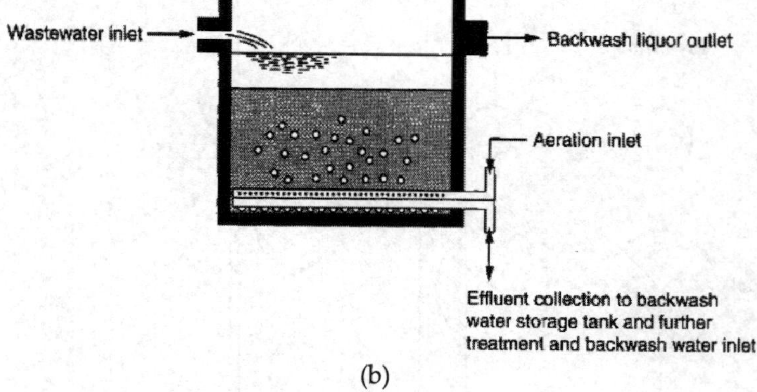

(b)

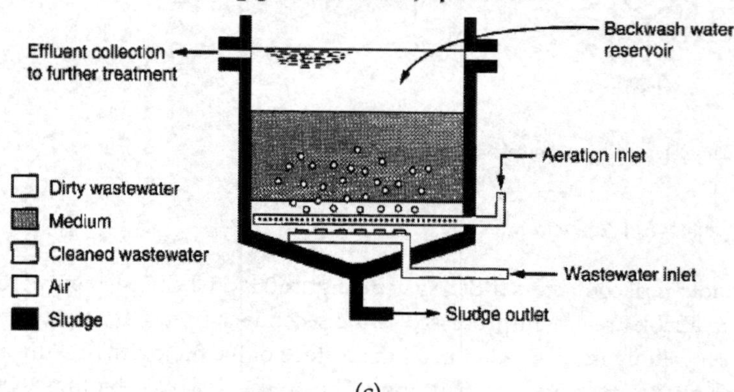

(c)

Figure 30. Biological aerated filters [1].

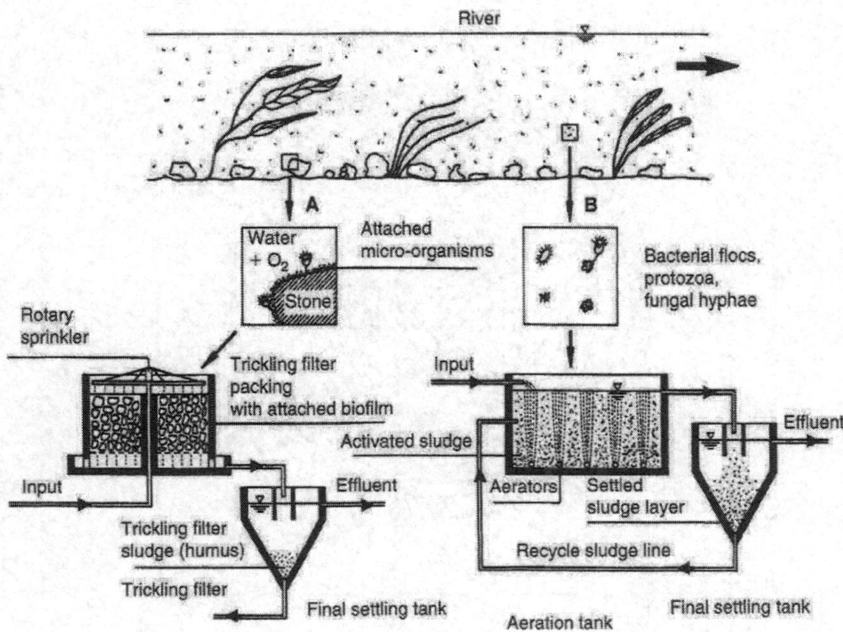

Figure 31. Relationship between the natural bacterial populations in rivers and the development of (A) trickling (percolating) filter and (B) activated sludge system [1].

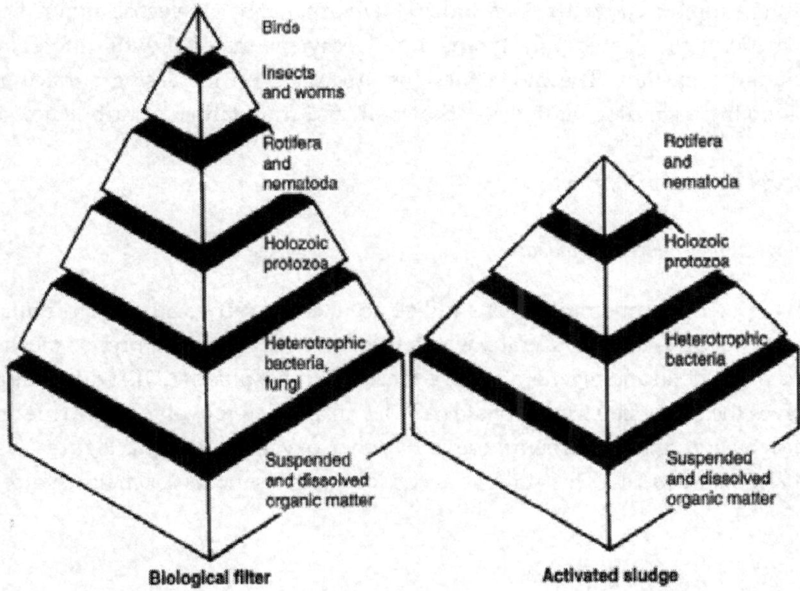

Figure 32. Comparison of the food chain pyramids for biological filters and activated sludge systems [1].

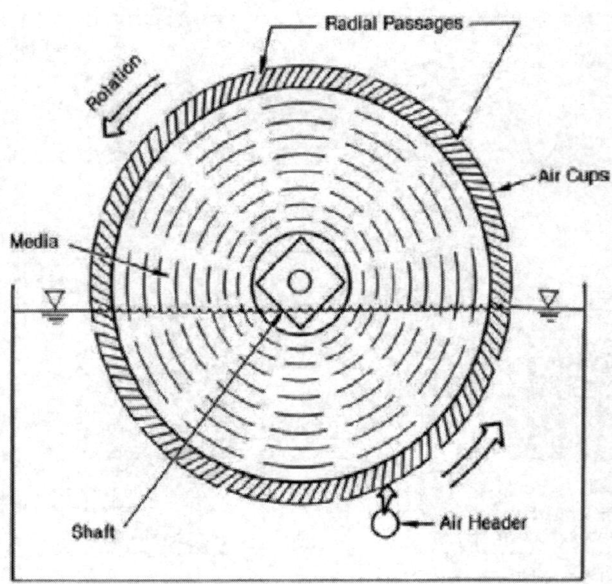

Figure 33. Schematic diagram of air-drive RBC [2].

The RBC consists of media, shaft, drive, bearings, and cover (Figure 34). The RBC hardware consists of a large diameter and closely spaced circular plastic media that is mounted on a horizontal shaft supported by bearings and is slowly rotated by an electric motor. The plastic media are made of corrugated polystyrene or polyethylene material with different designs, dimensions, and densities. The model designs are based on increasing surface area and firmness, allowing a winding wastewater flow path and stimulating air turbulence [1, 2].

4.4.7. Biological removal of nutrients

4.4.7.1. Biological phosphorous removal

It is widely agreed that microorganisms utilize acetate and fatty acids to accumulate poly-phosphates as poly-β-hydroxybutyrate, which is an acid polymer. The precise mechanism is based on the production and regeneration of adenosine diphosphate (ADP) within the bacteria, and it involves the adenosine triphosphate (ATP). Phosphate removal requires true anaerobic conditions, which occur only when there is no other oxygen donor [3]. Figure 36 shows a phosphate removal process. This process needs long narrow tanks for maintenance of plug flow.

4.4.7.2. Biological removal of nitrogen

The nitrification and denitrification processes are responsible for N_2O production (Figure 37). Figure 38 shows a nitrification/denitrification system for biological removal of nitrogen.

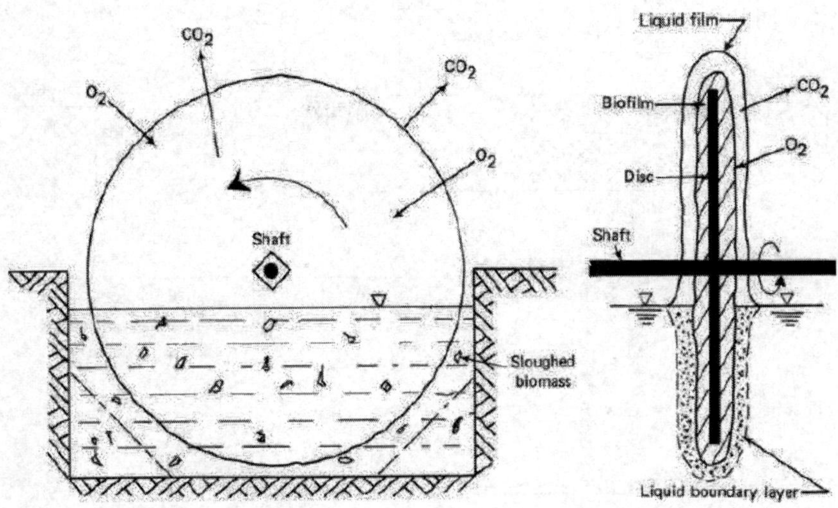

Figure 34. Mechanism of attached growth media in an RBC system [2].

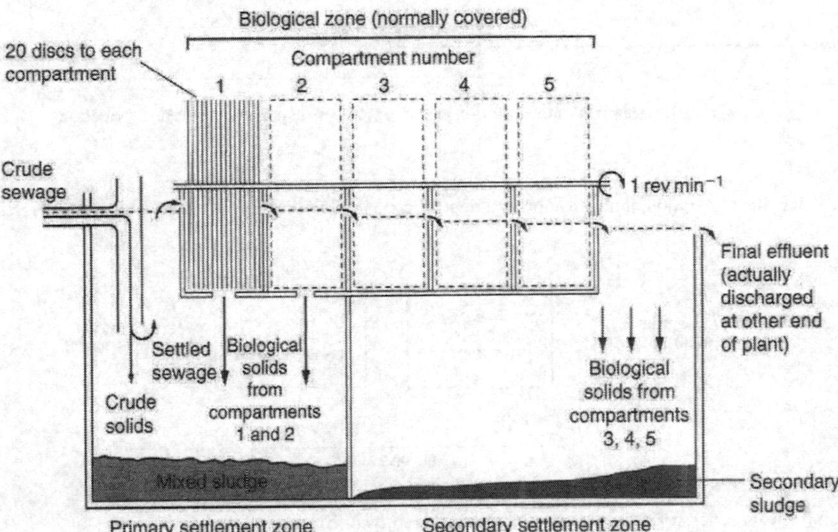

Figure 35. RBC system [1].

4.4.8. Phytoremediation

Phytoremediation is a treatment process that solves environmental problems by implementing plants that abate environmental pollution without excavating the pollutants and disposing them elsewhere. Phytoremediation is the abatement of pollutant concentrations in contaminated soils or water using plants that are able to accumulate, degrade, or eliminate heavy metals, pesticides, solvents, explosives, crude oils and its derivatives, and a multitude of other

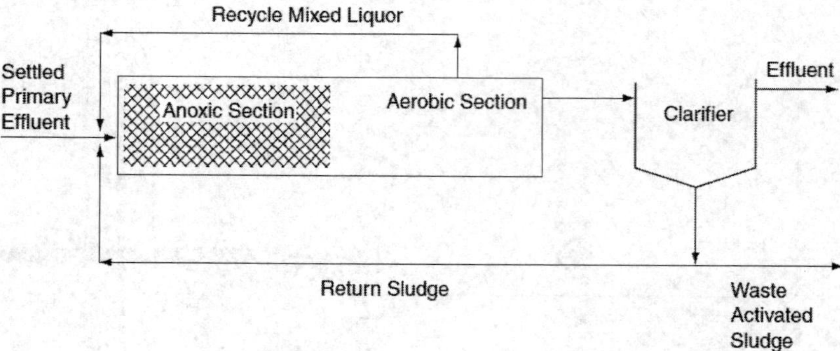

Figure 36. Phosphate removal process [3].

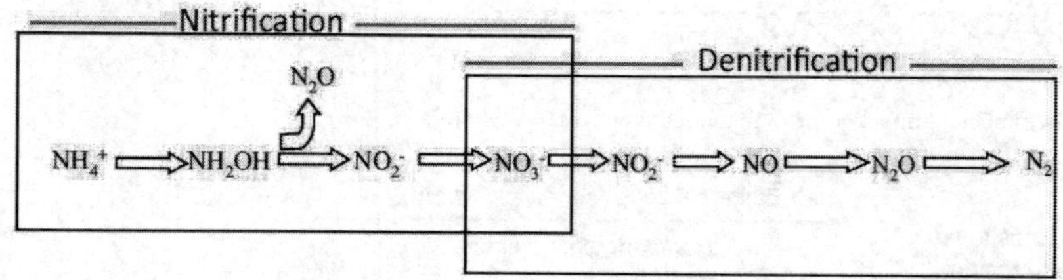

Figure 37. Schematic illustration of nitrification and denitrification processes that are responsible for N_2O release [16].

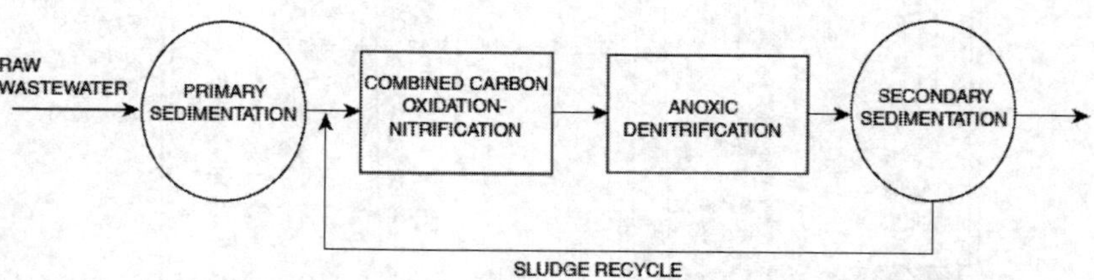

Figure 38. Nitrification/denitrification system for biological removal of nitrogen [3].

contaminants and pollutants from water and soils. Figures 39 through 44 show the designs of constructed wetlands where the phytoremediation takes place.

The incorporation of heavy metals, such as mercury, into the food chain may be a deteriorating matter. Phytoremediation is useful in these situations, where natural plants or transgenic plants are able to phytodegrade and phytoaccumulate these toxic contaminants in their above-ground parts, which will be then harvested for extraction. The heavy metals in the harvested biomass can be further concentrated by incineration and recycled for industrial implementation. Rhizofiltration is a sort of phytoremediation that involves filtering wastewater through

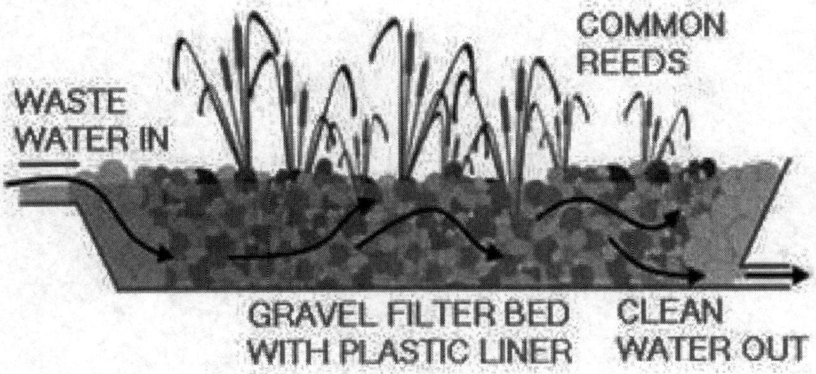

Figure 39. Cross-sectional view of a typical subsurface flow constructed wetland [17].

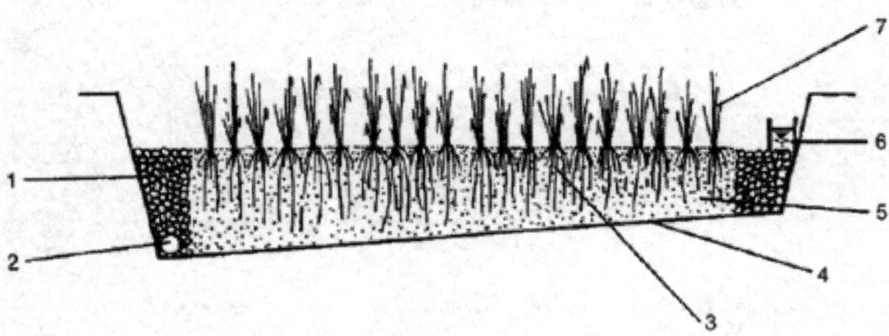

Figure 40. Components of a horizontal flow reed bed: (1) drainage zone consisting of large rocks, (2) drainage tube of treated effluent, (3) root zone, (4) impermeable liner, (5) soil or gravel, (6) wastewater distribution system, and (7) reeds [1].

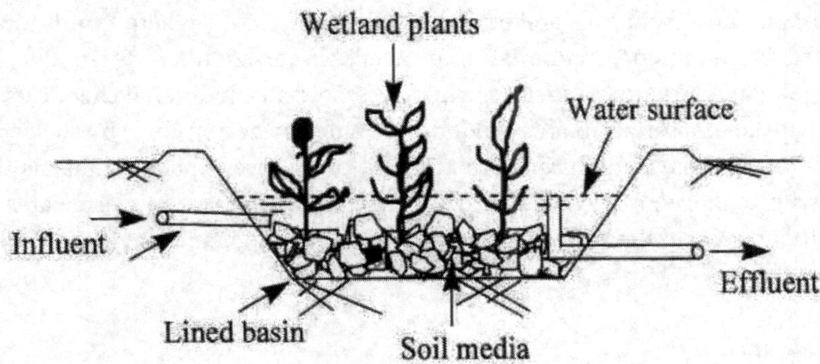

Figure 41. Free water surface system [18].

a mass of roots to remove toxic substances or excess nutrients. Phytoaccumulation or phytoextraction implements plants or algae to remove pollutants and contaminants from waste-

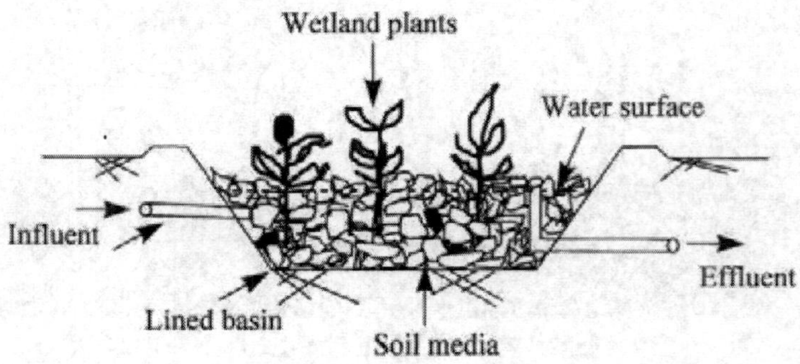

Figure 42. Sub-surface flow system [18].

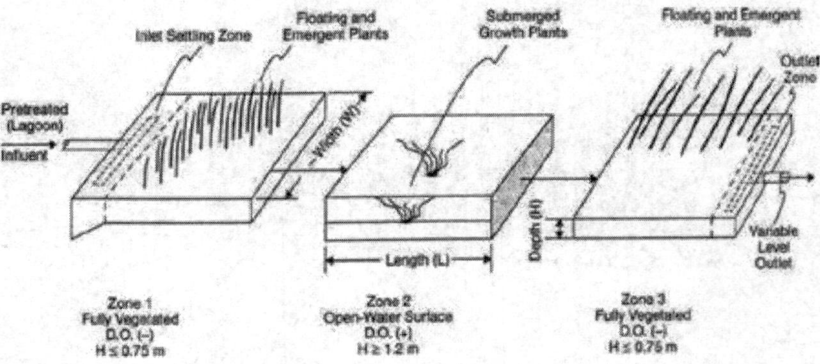

Figure 43. Components of a free water surface constructed wetland [2].

water into plant biomass that can be harvested. Organisms that accumulate over than usual amounts of pollutants from soils are termed hyperaccumulators, where a multitude of tables that show the different hyperaccumulators are available and should be referred to. In the case of organic pollutants, such as pesticides, explosives, solvents, industrial chemicals, and other xenobiotic substances, certain plants render these substances non-toxic by their metabolism and this process is called phytotransformation. In other cases, microorganisms that live in symbiosis with plant roots are able to metabolize these pollutants in wastewater. Figure 45 shows the tissues where the rhizofiltration, phytodegradation, and phytoaccumulation take place.

4.4.9. Vermifiltration

Vermiculture, or worm farming, is the implementation of some species of earthworm, such as *Eisenia fetida* (known as red wiggler, brandling, or manure worm) and *Lumbricus rubellus,* to make vermicompost, also known as worm compost, vermicast, worm castings, worm humus, or worm manure, which is the end-product of the breakdown of organic matter

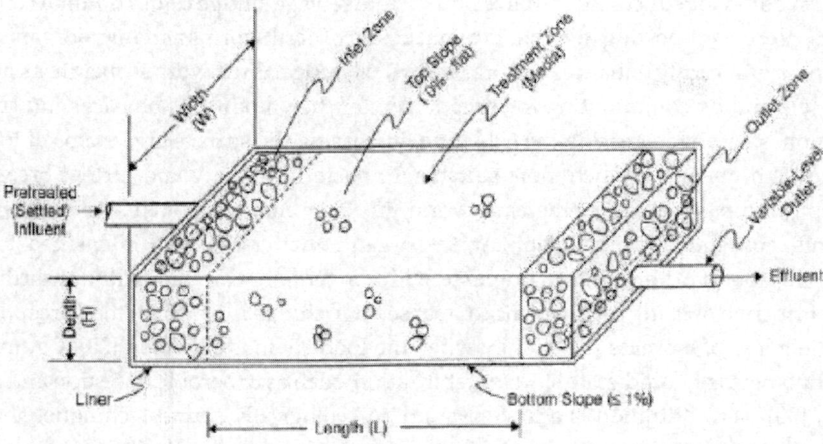

Figure 44. Components of a vegetated submerged bed system [2].

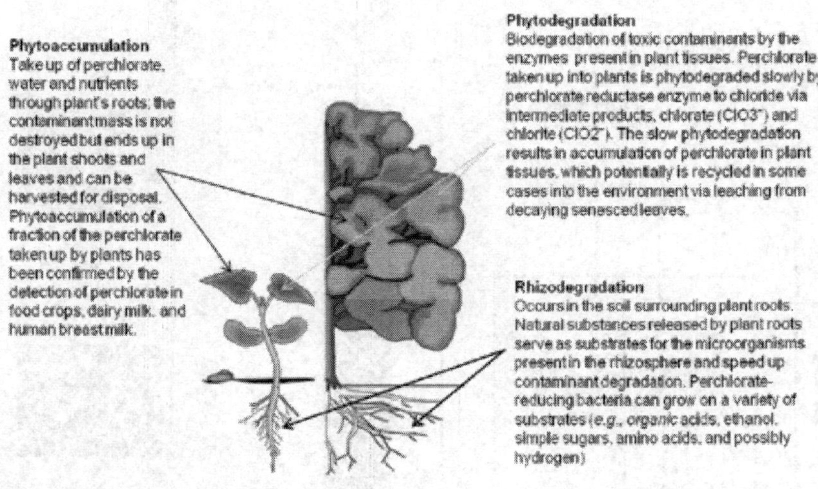

Figure 45. Rhizofiltration, phytodegradation, and phytoaccumulation [19].

and considered to be a nutrient-rich biofertilizer and soil conditioner. Vermiculture can be implemented to transform livestock manure, food leftovers, and organic matters into a nutrient-rich biofertilizer.

The potential use of earthworms to break down and manage sewage sludge began in the late 1970s [20] and was termed vermicomposting. The introduction of earthworms to the filtration systems, termed vermifiltration systems, was advocated by José Toha in 1992 [21]. Vermifilter is widely used to treat wastewater, and appeared to have high treatment efficiency, including synchronous stabilization of wastewater and sludge [22, 23, 24]. Vermifiltration is a feasible treatment method to reduce and stabilize liquid-state sewage sludge under optimal conditions [24, 25, 26]. Vermicomposting involves the joint action of earthworms and microorganisms [24, 27, 28], and significantly enhances the breakdown of sludge. Earthworms operate as mechanical blenders and by comminuting the organic matter they modify its physical and chemical composition, steadily decreasing the C:N ratio, increasing the surface area exposed to microorganisms, and making it much more suitable for bacterial activity and further breakdown. Throughout the passageway is the earthworm gut, they move fragments and bacteria-rich excrements, consequently homogenizing the organic matter [29]. An intensified bacterial diversity was found in vermifilter, compared with conventional biofilter without earthworms [25]. The principle of using earthworms to treat sewage sludge is based on the perception that there is a net loss of biomass and energy when the food chain is extended [25]. Compared to other technologies of liquid-state sludge stabilization, such as anaerobic digestion and aerobic digestion [30], vermifiltration is a low-cost and an ecologically sound technique, and more suitable for sewage sludge treatment of small or developing-countries' WWTPs [23, 24, 25, 26, 31]. Figure 46 illustrates schematic diagram of a vermifilter, where the earthworms are in the filter bed.

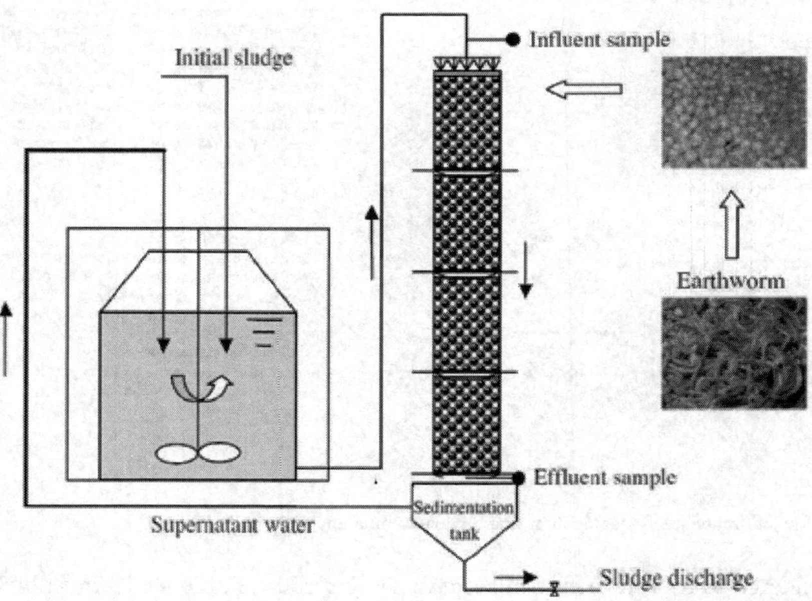

Figure 46. Schematic diagram of a vermifilter [24].

An important application is in livestock manure treatment as shown in Figure 47, where manure is flushed out from the livestock building to a raw effluent tank then the raw effluent is screened to separate the solid waste from manure. The screened effluent is then introduced to the vermifilter to produce the vermicompost. The vermifiltered effluent is then stored in a sedimentation tank. Afterwards, the vermifiltered effluent is introduced to constructed wetlands where the phytoremediation process takes place. The purified water can be then used to flush the water from the livestock building.

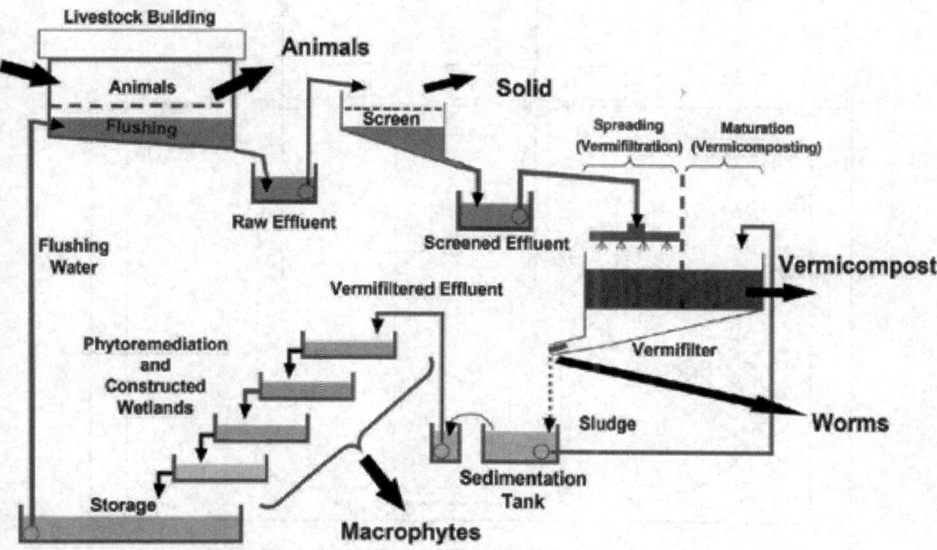

Figure 47. Schematic diagram of a manure treatment system containing vermifiltration and phytoremediation processes (Amended and redrawn from Morand et al. [32]).

4.4.10. Microbial fuel cells

The microbial fuel cells (MFCs) allow bacteria to grow on the anode by oxidizing the organic matter that result in releasing electrons. The cathode is sparked with air to provide dissolved oxygen for the reaction of electrons, protons, and oxygen on the cathode, which result in completing the electrical circuit and producing electrical energy (Figure 48).

5. Chemical treatment of wastewater

5.1. Chemical precipitation

The dissolved inorganic components can be removed by adding an acid or alkali, by changing the temperature, or by precipitation as a solid. The precipitate can be removed by sedimentation, flotation, or other solid removal processes [1]. Although chemical precipitation (coagulation, flocculation) is still implemented, it is highly recommended to substitute the chemical precipitation process by phytoremediation (see previous section), where the trend is to ramp up the implementation of bioremediation and phytoremediation to reduce the use of chemicals, which is in line with the "Green Development".

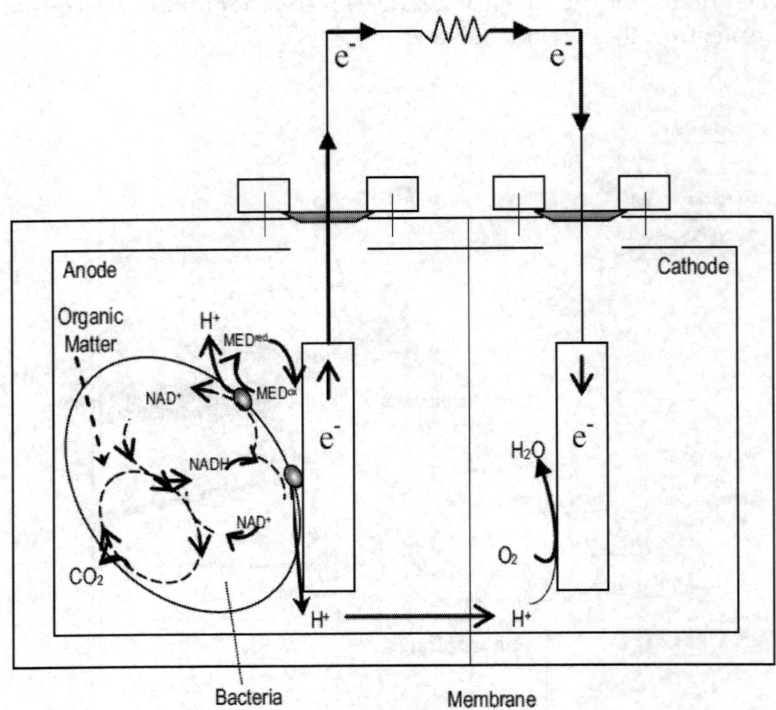

Figure 48. Schematic diagram of the essential components of an MFC [33].

5.2. Neutralization

Neutralization is controlling the pH of the wastewater whether it is acidic or alkaline to keep the pH around 7. The lack of sufficient alkalinity will require the addition of a base (Table 3) to adjust the pH to the acceptable range. Lime (CaO), calcium hydroxide ($Ca(OH)_2$), sodium hydroxide (NaOH), and sodium carbonate (Na_2CO_3), also known as soda ash, are the most common chemicals used to adjust the pH [34]. The lack of sufficient acidity will require the addition of an acid to adjust the pH to the acceptable range. Sulfuric acid (H_2SO_4) and carbonic acid (H_2CO_3) are the most common chemicals used to adjust the pH.

5.3. Adsorption

Adsorption is a physical process where soluble molecules (adsorbate) are removed by attachment to the surface of a solid substrate (adsorbent). Adsorbents should have an extremely high specific surface area. Examples of adsorbents include activated alumina, clay colloids, hydroxides, resins, and activated carbon. The surface of the adsorbent should be free of adsorbate. Therefore, the adsorbent should be activated before use. A wide range of organic materials can be removed by adsorption, including detergents and toxic compounds. The most widely used adsorbent is activated carbon, which can be produced by pyrolytic carbonization of biomass [1]. Figure 49 illustrates the difference between absorption and adsorption.

Neutralization reactions

To neutralize sulfuric acid with	
Lime:	$H_2SO_4 + CaO \rightleftharpoons CaSO_4 + H_2O$
Calcium hydroxide:	$H_2SO_4 + Ca(OH)_2 \rightleftharpoons CaSO_4 + H_2O$
Sodium hydroxide:	$H_2SO_4 + NaOH \rightleftharpoons NaSO_4 + 2H_2O$
Soda ash:	$H_2SO_4 + Na_2CO_3 \rightleftharpoons Na_2SO_4 + H_2O + CO_2$
To neutralize hydrochloric acid with	
Lime:	$2HCl + CaO \rightleftharpoons CaCl_2 + H_2O$
Calcium hydroxide:	$HCl + Ca(OH)_2 \rightleftharpoons CaCl_2 + H_2O$
Sodium hydroxide:	$HCl + NaOH \rightleftharpoons NaCl + H_2O$
Soda ash:	$HCl + Na_2CO_3 \rightleftharpoons NaCl + H_2O + CO_2$

Note: a stoichiometric reaction will yield a pH of 7.0.

Note: a stoichiometric reaction will yield a pH of 7.0

Table 3. Neutralization: Case of acidic wastewater [34].

Activated carbon is the most implemented adsorbent and is a sort of carbon processed to be riddled with small, low-volume pores that enlarge the surface area available for adsorption. Owing to its high level of microporosity, 1 g of activated carbon has a surface area larger than 500 m^2, which was determined by gas adsorption. Figure 50 shows a bed carbon adsorption unit. Note that the carbon can be regenerated by thermal oxidation or steam oxidation and reused. The adsorption capacity, one of the most important characteristics of an adsorbent, can be calculated as follows:

$$\underset{mg/g}{\text{Adsorption Capacity}} = \underset{mg}{\text{Adsorbate}} / \underset{g}{\text{Adsorbent}}$$

The factors that affect adsorption are [3]:

1. Particle diameter: the adsorption is inversely proportional to the particle size of the adsorbent, and directly proportional to surface area.

2. Adsorbate concentration: the adsorption is directly proportional to adsorbate concentration.

3. Temperature: the adsorption is directly proportional to temperature.

4. Molecular weight: generally, the adsorption is inversely proportional to molecular weight depending upon the compound weight and configuration of pores diffusion control.

5. pH: the adsorption is inversely proportional to pH due to surface charge.

6. Individual properties of adsorbate and adsorbent are difficult to compare.

7. Iodine number: is the mass of iodine (g) that is consumed by 100 g of a substance.

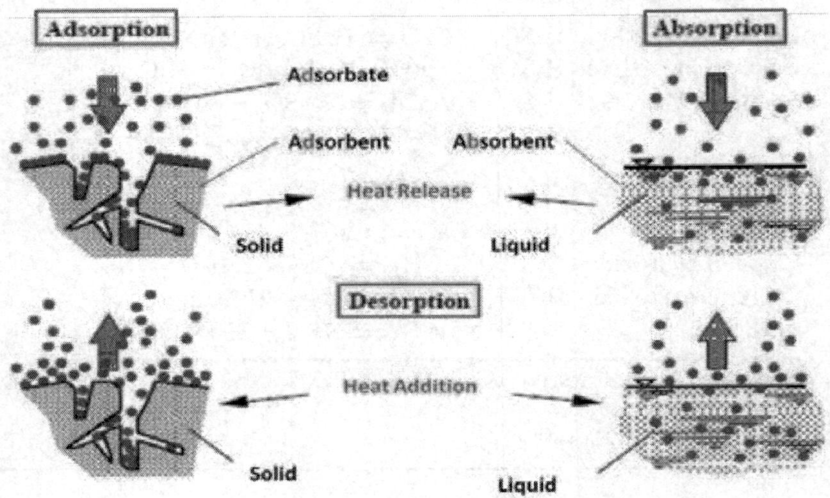

Figure 49. A comparison between absorption and adsorption.

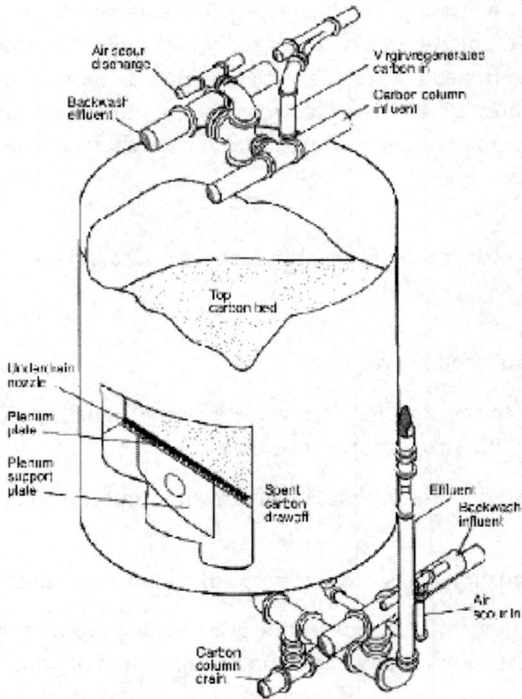

Figure 50. A bed carbon adsorption unit [35].

5.4. Disinfection

The disinfection of wastewater is the last treatment step of the tertiary treatment process. Disinfection is a chemical treatment process conducted by treating the effluent with the selected disinfectant to exterminate or at least inactivate the pathogens. The rationales behind effluent disinfection are to protect public health by exterminating or inactivating the pathogens such as microbes, viruses, and protozoan, and to meet the wastewater discharge standards. The purpose of disinfection is the protection of the microbial wastewater quality. The ideal disinfectant should have bacterial toxicity, is inexpensive, not dangerous to handle, and should have reliable means of detecting the presence of a residual. The chemical disinfection agents include chlorine, ozone, ultraviolet radiation, chlorine dioxide, and bromine [3].

5.4.1. Chlorine

Chlorine is one of the oldest disinfection agents used, which is one of the safest and most reliable. It has extremely good properties, which conform to the aspects of the ideal disinfectant. Effective chlorine disinfection depends upon its chemical form in wastewater. The influencing factors are pH, temperature, and organic content in the wastewater [3]. When chlorine gas is dissolved in wastewater, it rapidly hydrolyzes to hydrochloric acid (HCl) and hypochlorous acid (HOCl) as shown in the following chemical equation:

$$Cl_2 + H_2O \leftrightarrow H^+ + Cl^- + HOCl$$

Free ammonia combines with the HOCl form of chlorine to form chloramines in a three-step reaction, as follows:

$$NH_3 + HOCl \rightarrow NH_2Cl + H_2O$$
$$NH_2Cl + HOCl \rightarrow NHCl_2 + H_2O$$
$$NHCL_2 + HOCl \rightarrow NCl_3 + H_2O$$

Figure 51 illustrates the chlorination curve, where the formation of chloramines occurs at the breakpoint. The free chlorine residual first rises then falls until the reaction with ammonia has been completed. As additional chlorine is applied and ammonia is consumed, the chlorine residual rises again.

Dechlorination is a very important process, where activated carbon, sulfur compounds, hydrogen sulfide, and ammonia can be implemented to minimize the residual chlorine in a disinfected effluent prior to discharge. Activated carbon and sulfur compounds are the most widely used [3]. The commonly used sulfur compounds are sulfur dioxide (SO_2), sodium metabisulfite (NaS_2O_5), sodium bisulfate ($NaHSO_3$), and sodium sulfite (Na_2SO_3). The dechlorination reactions with the abovementioned compounds are described in the following equations:

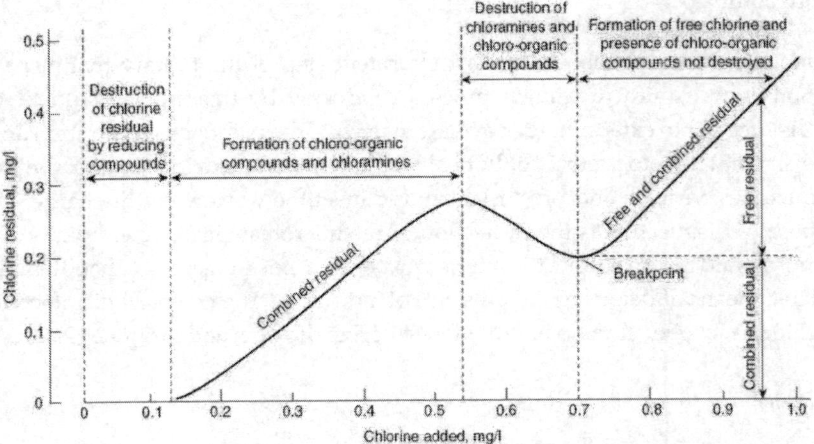

Break point chlorination by the formation of chloramines. The free chlorine residuals first rises then falls until the reaction with ammonia have been completed. As additional chlorine is applied and ammonia is consumed, the chlorine residual again rises.

Figure 51. Chlorination curve [3].

$$SO_2 + 2H_2O + Cl_2 \rightarrow H_2SO_4 + 2HCl$$
$$SO_2 + H_2O + HOCl \rightarrow 3H^+ + Cl^- + SO_4^{2-}$$
$$Na_2S_2O_5 + 2Cl_2 + 3H_2O \rightarrow 2NaHSO_4 + 4HCl$$
$$NaHSO_3 + H_2O + Cl_2 \rightarrow NaHSO_4 + 2HCl$$

5.4.2. Ozone

Ozone (O_3) is a very strong oxidant typically used in wastewater treatment. Ozone is able to oxidize a multitude of organic and inorganic compounds in wastewater. These reactions cause an ozone demand in the treated wastewater, which should be fulfilled throughout wastewater ozonation prior to developing an assessable residual. Ozone should be generated at the point of application for use in wastewater treatment as ozone is an unstable molecule [3]. Figure 52 illustrates the corona discharge method for making ozone. Ozone is generally formed by combining an oxygen atom with an oxygen molecule (O_2) as follows:

5.4.3. Ultraviolet light

Ultraviolet (UV) radiation is a microbial disinfectant that leaves no residual. It requires clear, un-turbid, and non-colored water for its implementation. The commercial UV disinfection systems use low- to medium-powered UV lamps with a wavelength of 354 nm [3]. The UV dosage can be calculated as follows:

$$D = I \cdot t$$

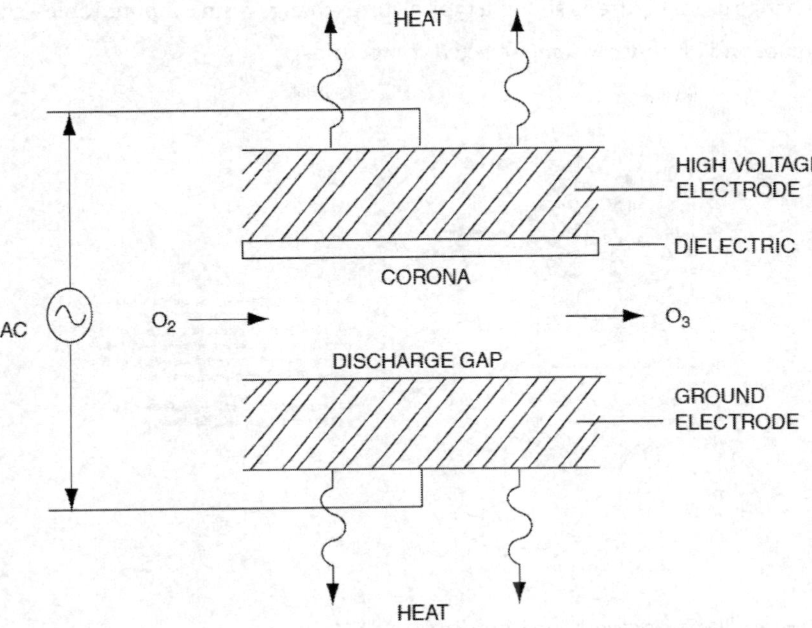

Figure 52. Schematic drawing of corona discharge method for making ozone [3].

where, D is the UV dose (mW. s/cm^2); I is the intensity (mW/cm^2); and t is the exposure time (s).

The advantages of UV radiation are: (1) directly effective against the DNA of many microorganisms, (2) not reactive with other forms of carbonaceous demand, and (3) provides superior bactericidal kill values while not leaving any residues. The advantage is often the disadvantage, because power fluctuations, variations in hydraulic flow rates, and color or turbidity can cause the treatment to be ineffective [3]. Additionally, cell recovery and re-growth of the damaged organisms because of the inactivation of their predators and competitors has come to light.

5.5. Ion exchange

Ion exchange (IX) is a reversible reaction in which a charged ion in a solution is exchanged with a similarly charged ion which is electrostatically attached to an immobile solid particle. The most common implementation of ion exchange method in wastewater treatment is for softening, where polyvalent cations (e.g., calcium and magnesium) are exchanged with sodium [36]. Practically, wastewater is introduced into a bed of resin. The resin is manufactured by converting a polymerization of organic compounds into a porous matrix. Typically, sodium is exchanged with cations in the solution [34]. The bed is shut down when it becomes saturated with the exchanged ions, where it should be regenerated by passing a concentrated solution of sodium back through the bed. Figure 53 shows the schematic illustration of organic cation-exchange bead. Figure 54 shows a typical ion exchange resin column. Table 4 shows the ion preference and affinity for some selected compounds.

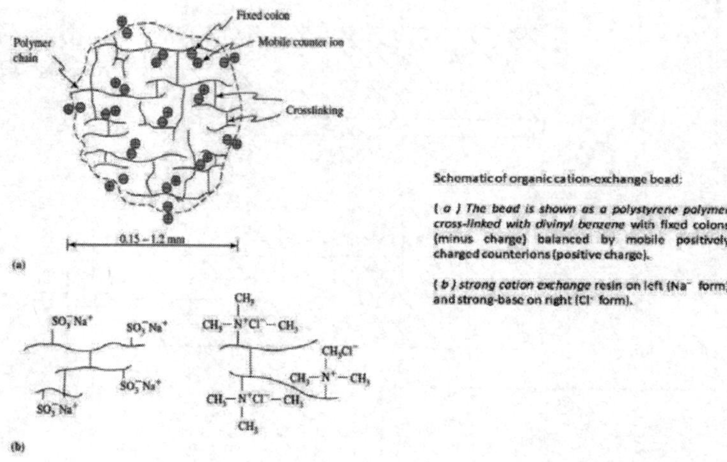

Figure 53. Schematic illustration of organic cation-exchange bead [34].

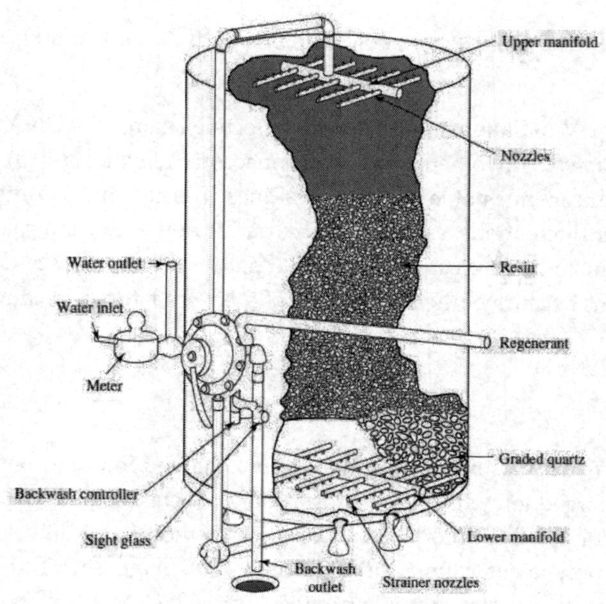

Figure 54. Typical ion exchange resin column [37].

Strong Acid Cation Exchanger	Strong Base Anion Exchanger	Weak Acid Cation Exchanger	Weak Base Anion Exchanger	Weak Acid Chelate Exchanger
Barium (2+)	Iodide (1−)	Hydrogen (1+)	Hydroxide (1−)	Copper (2+)
Lead (2+)	Nitrate (1−)	Copper (2+)	Sulfate (2−)	Iron (2+)
Mercury (2+)	Bisulfite (1−)	Cobalt (2+)	Chromate (2−)	Nickel (2+)
Copper (1+)	Chloride (1−)	Nickel (2+)	Phosphate (2−)	Lead (2+)
Calcium (2+)	Cyanide (1−)	Calcium (2+)	Chloride (1−)	Manganese (2+)
Nickel (2+)	Bicarbonate (1−)	Magnesium (2+)		Calcium (2+)
Cadmium (2+)	Hydroxide (1−)	Sodium (1+)		Magnesium (2+)
Copper (2+)	Fluoride (1−)			Sodium (1+)
Cobalt (2+)	Sulfate (2−)			
Zinc (2+)				
Cesium (1+)				
Iron (2+)				
Magnesium (2+)				
Potassium (1+)				
Manganese (2+)				
Ammonia (1+)				
Sodium (1+)				
Hydrogen (1+)				
Lithium (1+)				

Table 4. Ion preference and affinity for some selected compounds [3].

5.6. Physicochemical treatment processes

The principal advanced physicochemical wastewater treatment processes are elucidated in Table 5.

Principal advanced physico-chemical wastewater treatment processes

Process	*Removal function*
Filtration	Suspended solids
Air stripping	Ammonia
Breakpoint chlorination	Ammonia
Ion exchange	Nitrate, dissolved inorganic solids
Chemical precipitation	Phosphorus, dissolved inorganic solids
Carbon adsorption	Toxic compounds, refractory organics
Chemical oxidation	Toxic compounds, refractory organics
Ultrafiltration	Dissolved inorganic solids
Reverse osmosis	Dissolved inorganic solids
Electrodialysis	Dissolved inorganic solids
Volatilization and gas stripping	Volatile organic compounds

Table 5. Principal advanced physicochemical wastewater treatment processes [1].

6. Wastewater treatment plants

This section shows some examples of WWTPs as shown in Figure 55 (a, b, and c) and Figure 56. On the other hand, there are some computer programs for planning and designing WWTPs (Figures 57, 58, and 59).

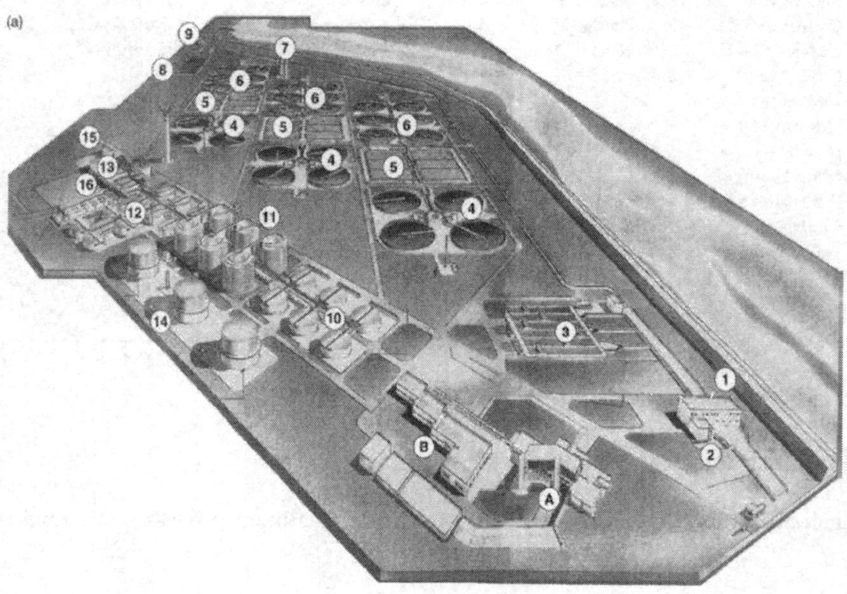

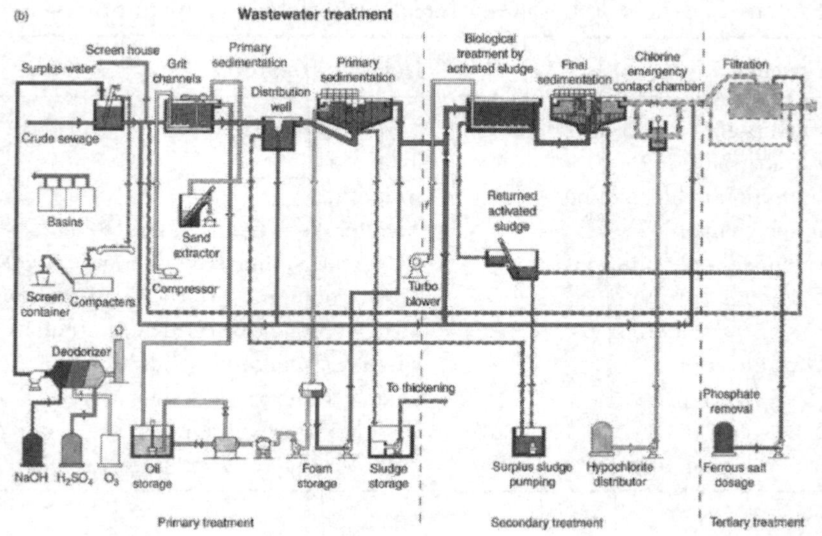

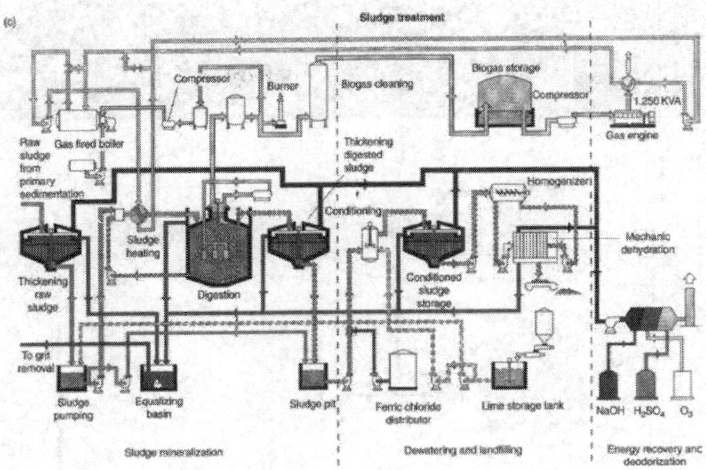

Figure 55. WWTP showing: (a) layout of the plant, (b) wastewater process flow diagrams, and (c) sludge process flow diagram. **Wastewater treatment**: 1. Storm water overflow; 2. screening; 3. grit removal; 4. primary sedimentation; 5. aeration tanks; 6. Secondary sedimentation; 7. emergency chlorination; 8. filtration; 9. effluent outfall. **Sludge treatment**: 10. raw sludge thickeners; 11. digestion tanks; 12. digested sludge thickeners; 13. power house; 14. biogas storage; 15. filter press house; 16. transformer station. A and B are administrative areas [1].

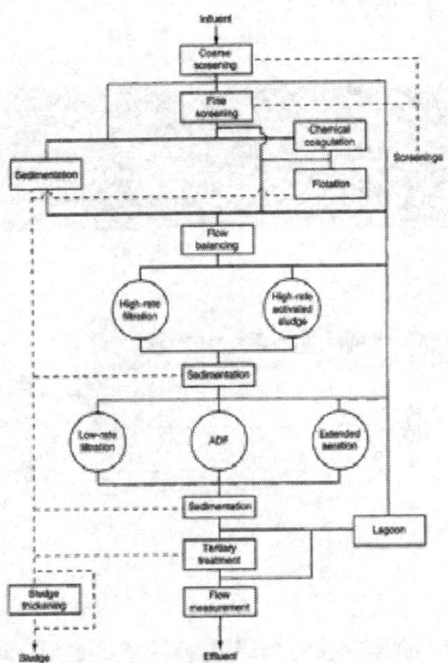

Figure 56. Summary of the main process options commonly employed at both domestic and industrial WWTPs. Not all of these unit processes may be selected, but the order of their use remains the same [1].

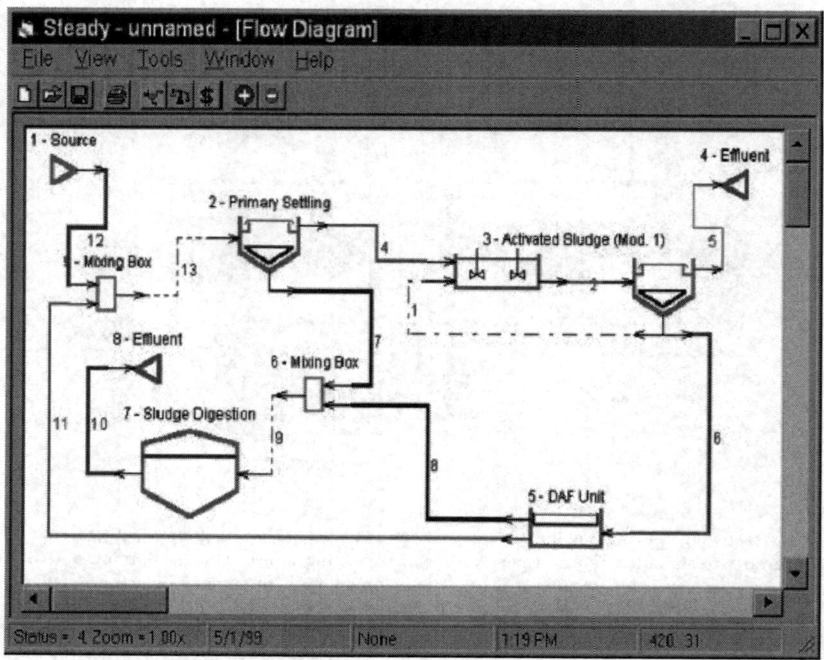

Figure 57. Screenshot of the STEADY program [3].

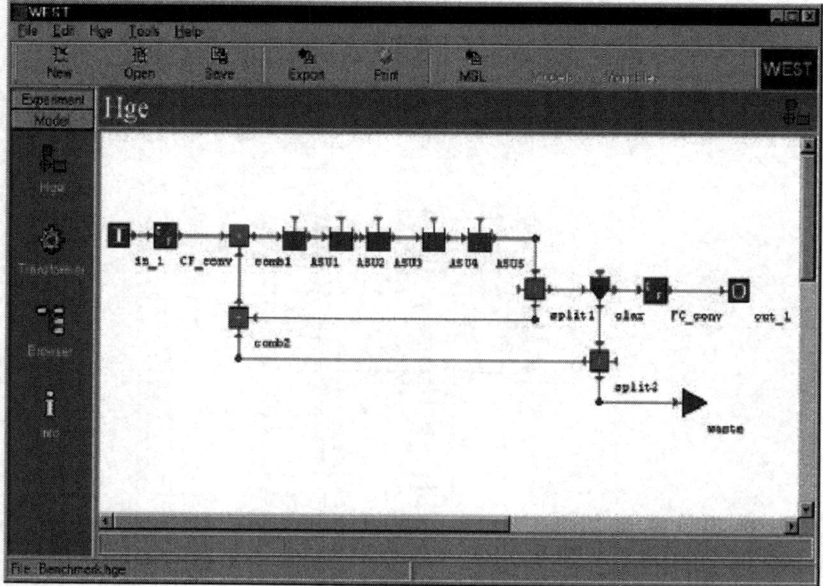

Figure 58. WEST software typical plant configuration [3].

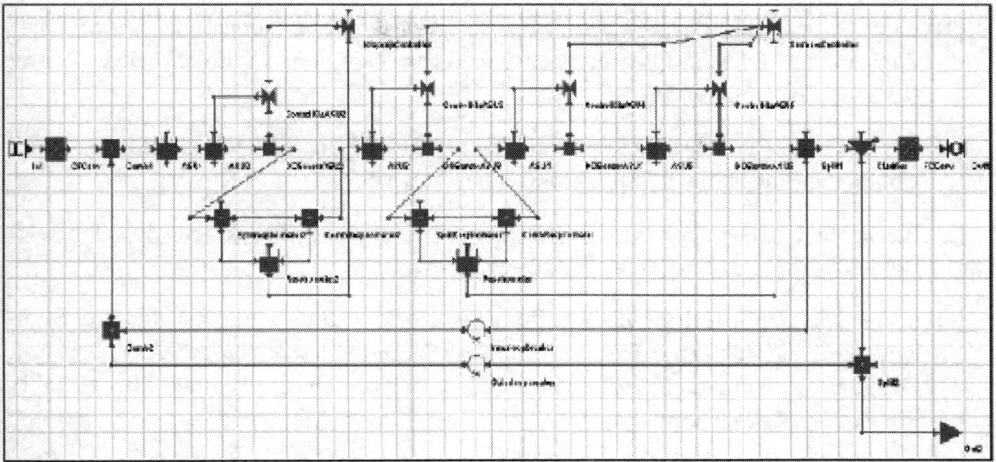

Figure 59. WEST configuration for multitank system [3].

7. Conclusions

According to this study, it can be conclude that:

1. The trend is to ramp up the implementation of bioremediation, phytoremediation, and mycoremediation to reduce the use of chemicals, which is in line with the "Green Development".
2. The recent developments elucidate that subsequent to the physical treatment processes (the primary treatment) the biological treatment processes come in turn as secondary treatment and precede the chemical treatment processes, which constitute the tertiary treatment.
3. Microbial fuel cells, phytoremediation, and mycoremediation are the focus of the future development in this field.

References

[1] Gray N. F. (2005). *Water Technology: An Introduction for Environmental Scientists and Engineers (2nd Edition)*, Elsevier Science & Technology Books, ISBN 0750666331, Amsterdam, The Netherlands.

[2] Lin, S. D. (2007). *Water and Wastewater Calculations Manual (2nd Edition)*, McGraw-Hill Companies, Inc., ISBN 0-07-154266-3, New York, USA.

[3] Russell D. L. (2006). *Practical Wastewater Treatment*, John Wiley & Sons, Inc., ISBN-13: 978-0-471-78044-1, Hoboken, New Jersey, USA.

[4] Samer M. (2015). GHG Emission from Livestock Manure and its Mitigation Strategies, In: *Climate Change Impact on Livestock: Adaptation and Mitigation*, V. Sejian, J. Gaughan, L. Baumgard & C. Prasad (Eds.), In Press, Springer International, ISBN 978-81-322-2264-4, Germany.

[5] Samer M., Mostafa E. & Hassan A. M. (2014). Slurry Treatment with Food Industry Wastes for Reducing Methane, Nitrous Oxide and Ammonia Emissions. *Misr Journal of Agricultural Engineering*, 31 (4): 1523–1548.

[6] Samer M. (2011). How to Construct Manure Storages and Handling Systems? *IST Transactions of Biosystems and Agricultural Engineering*, Vol. 1, No. 1, pp. 1–7, ISSN 1913-8741.

[7] Samer M., Grimm H., Hatem M., Doluschitz R. & Jungbluth T. (2008). Mathematical Modeling and Spark Mapping for Construction of Aerobic Treatment Systems and their Manure Handling System, *Proceedings of International Conference on Agricultural Engineering*, Book of Abstracts p. 28, EurAgEng, Crete, Greece, June 23-25, 2008.

[8] Ersahin M. E., Ozgun H., Dereli R. K. & Ozturk I. (2011). Anaerobic Treatment of Industrial Effluents: An Overview of Applications, In: *Waste Water - Treatment and Reutilization*, F. Einschlag (Ed.), pp. 3–28, InTech, ISBN 978-978-953-307-249-4, Rijeka, Croatia.

[9] Samer M. (2012). Biogas Plant Constructions, In: *Biogas*, S. Kumar (Ed.), pp. 343–368, InTech, ISBN 978-953-51-0204-5, Rijeka, Croatia.

[10] Samer M. (2010). A Software Program for Planning and Designing Biogas Plants. *Transactions of the ASABE*, Vol. 53, No. 4, pp. 1277–1285, ISSN 2151-0032.

[11] Pfost D. L. & Fulhage C. D. (2000). Anaerobic Lagoons for Storage/Treatment of Livestock Manure, EQ 387, MU Extension University of Missouri-Columbia, USA.

[12] Westerman P. W., Shaffer K. A. & Rice J. M. (2008). Sludge Survey Methods for Anaerobic Lagoons, AG-639W, North Carolina Cooperative Extension Service, College of Agriculture & Life Sciences, North Carolina State University, USA.

[13] Pavko A. (2011). Fungal Decolourization and Degradation of Synthetic Dyes Some Chemical Engineering Aspects, In: *Waste Water - Treatment and Reutilization*, F. Einschlag (Ed.), pp. 65–88, InTech, ISBN 978-978-953-307-249-4, Rijeka, Croatia.

[14] Diano N. & Mita D.G. (2011). Removal of Endocrine Disruptors in Waste Waters by Means of Bioreactors, In: *Waste Water - Treatment and Reutilization*, F. Einschlag (Ed.), pp. 29-48, InTech, ISBN 978-978-953-307-249-4, Rijeka, Croatia.

[15] Kabir M., Suzuki M. & Yoshimura N. (2011). Excess Sludge Reduction in Waste Water Treatment Plants, In: *Waste Water - Treatment and Reutilization*, F. Einschlag (Ed.), pp. 133–150, InTech, ISBN 978-978-953-307-249-4, Rijeka, Croatia.

[16] Chadwick D., Sommer S., Thorman R., Fangueiro D., Cardenas L., Amon B. & Misselbrook T. (2011). Manure Management: Implications for Greenhouse Gas Emissions. *Animal Feed Science and Technology*, 166–167(2011): 514– 531, ISSN 0377-8401.

[17] Morris R. H. & Knowles P. (2011). Measurement Techniques for Wastewater Filtration Systems, In: *Waste Water - Treatment and Reutilization*, F. Einschlag (Ed.), pp. 109–132, InTech, ISBN 978-978-953-307-249-4, Rijeka, Croatia

[18] ESCWA (2003). Economic and Social Commission for Western Asia.

[19] Kucharzyk K. H., Soule T., Paszczynski A. J. & Hess T. F. (2011). Perchlorate: Status and Overview of New Remedial Technologies, In: *Waste Water - Treatment and Reutilization*, F. Einschlag (Ed.), pp. 171–194, InTech, ISBN 978-978-953-307-249-4, Rijeka, Croatia.

[20] Li, X., Xing M., Yang J. & Huang Z. (2011). Compositional and functional features of humic acid-like fractions from vermicomposting of sewage sludge and cow dung, *J. Hazard. Mater.* 185: 740–748.

[21] Li Y., Robin P., Cluzeau D., Bouché M., Qiu J., Laplanche A., Hassouna M., Morand P., Dappelo C. & Callarec J. (2008). Vermifiltration as a stage in reuse of swine wastewater: monitoring methodology on an experimental farm, *Ecol. Eng.* 32: 301–309.

[22] Li Y., Xiao Y., Qiu J., Dai Y. & Robin P. (2009). Continuous village sewage treatment by vermifiltration and activated sludge process, *Water Sci. Technol.* 60: 3001–3010.

[23] Xing M., Yang J., Wang Y., Liu J. & Yu F. (2011). A comparative study of synchronous treatment of sewage and sludge by two vermifiltration using epigeic earthworm Eisenia fetida, *J. Hazard. Mater.* 185 (2–3): 881–888.

[24] Xing M., Li X., Yang J., Lv B. & Lu Y. (2012). Performance and mechanism of vermifiltration system for liquid-state sewage sludge treatment using molecular and stable isotopic techniques. *Chemical Engineering Journal* 197: 143–150.

[25] Zhao L., Wang Y., Yang J., Xing M., Li X., Yi D. & Deng D. (2010). Earthworm-microorganism interactions: a strategy to stabilize domestic wastewater sludge, *Water Res.* 44: 2572–2582.

[26] Xing M., Zhao L., Yang J., Huang Z. & Xu Z. (2011). Distribution and transformation of organic matter during liquid-state vermiconversion of activated sludge using elemental analysis and spectroscopic evaluation, *Environ. Eng. Sci.* 28 (9): 619–626.

[27] Suthar S. (2008). Development of a novel epigeic–anecic-based polyculture vermireactor for efficient treatment of municipal sewage water sludge, *Int. J. Environ. Waster Manage.* 2: 84–101.

[28] Xing M., Li X., Yang J., Huang Z. & Lu Y. (2011). Changes in the chemical characteristics of water-extracted organic matter from vermicomposting of sewage sludge and cow dung, *J. Hazard. Mater.* http://dx.doi.org/10.1016/j.jhazmat.2011.11.070.

[29] Domínguez J. & Edwards C.A. (2004). Vermicomposting organic wastes, in: S.H. Shakir Hanna, W.Z.A. Mikhaïl (Eds.), *Soil Zoology for Sustainable Development in the 21st Century,* Cairo, pp. 369–395.

[30] Fytili D. & Zabaniotou A. (2008). Utilization of sewage sludge in EU application of old and new methods – a review, *Renew. Sust. Energ. Rev.* 12: 116–140.

[31] Sinha R.K., Bharambe G. & Chaudhari U. (2008). Sewage treatment by vermifiltration with synchronous treatment of sludge by earthworms: A low-cost sustainable technology over conventional systems with potential for decentralization, *Environmentalist.* 28: 409–420.

[32] Morand P., Robin P., Qiu J.P., Li Y., Cluzeau D., Hamon G., Amblard C., Fievet S., Oudart D., Pain le Quere C., Pourcher A.-M., Escande A., Picot B. & Landrain B. (2009). Biomass production and water purification from fresh liquid manure by vermiculture, macrophytes ponds and constructed wetlands to recover nutrients and recycle water for flushing in pig housing. Proceedings of the International Congress "Ecological engineering: From concepts to applications Foreword (EECA)", 2-4 December 2009, Paris, France.

[33] Logan B. (2008). *Microbial Fuel Cells,* Wiley. Hoboken, NJ, EEUU. 200 pp.

[34] Davis M. L. (2010). *Water and Wastewater Engineering: Design Principles and Practice,* McGraw-Hill Companies, Inc., ISBN 978-0-07-171385-6, New York, USA.

[35] Sincero, A. P. & G. A. Sincero (2003). *Physical-Chemical Treatment of Water and Wastewater,* CRC Press LLC, ISBN 1-84339-028-0, Boca Raton, Florida, USA.

[36] Clifford D. A. (1999). Ion Exchange and Inorganic Adsorption, In: *Water Quality and Treatment (5th Edition),* R. D. Letterman (Ed.), pp. 9.1–9.91., American Water Works Association, McGraw-Hill, New York, USA.

[37] U.S. EPA (1981) Development Document for Effluent Limitations: Guideline and Standards for the Metal Finishing Point Source Category, U.S. Environmental Protection Agency, EPA/440/1-83-091, Washington, D.C., USA.

[38] U.S. EPA (2002). Wastewater Technology Fact Sheet: Anaerobic Lagoons, U.S. Environmental Protection Agency, EPA 832-F-02-009, Washington, D.C., USA.

Chapter 2

Bioremediation of Nitroaromatic Compounds

Abstract

Nitroaromatics are major pollutants released in the environment during the post-industrialization era and pose toxic effects to living organisms. Several bacterial strains have been isolated for the degradation of these nitroaromatic pollutants. Some of them have been used in field trial experiments for the removal of nitroaromatics from industrial water and groundwater. Very few bacterial pathways have been characterized at genetic and molecular levels. In this review, we cover all reported degradation pathways and their gene evolution. These studies for nitroaromatics clearly indicate that most of the involved genes have evolved from preexisting enzymes by using all means of gene evolution like horizontal gene transfer, mutation, and promiscuity principle. This information has been exploited for the creation of hybrid pathways and better biocatalysts for degradation.

Keywords: Nitroaromatics, biodegradation, gene evolution, oxygenase

1. Introduction

Microbes are empowered to degrade environmental pollutants under different conditions to perform unusual metabolic and physiological activities [1]. The metabolic versatility of the microbes helps them utilize a range of organic/inorganic compounds for their growth. This metabolic versatility has been exploited for human benefit in various industrial products like cheese (dairy industry), insulin, antibiotics (pharmaceutical industry), etc. Microbial enzyme systems have also been used for the development of suitable biocatalysts for green chemistry applications [2]. Whole microbes have also been tested for their potential in bioremediation [3]. Organic aromatic compounds have been the main source of pollution in several water bodies [4]. Thus, a remediation and degradation study of aromatics has been a focus of intensive study. In this review, we will focus only on nitroaromatic compounds.

1.1. Nitroaromatics

Nitroaromatic compounds have at least one nitro group attached to the aromatic ring, like nitrobenzene, nitrotoluenes, nitrophenols, etc. In nature, nitroaromatic compounds are mostly found in natural products from different plants, fungi, and bacteria [5, 6]. The best known example of this is chloramphenicol, which is produced by *Streptomyces venezuelae* [7–9]. The role of some nitroaromatic compounds in cellular signaling has also been established. For example, 2-nitrophenol and 4-methyl-2-nitrophenol are well-known pheromones for ticks that enable them to aggregate and attach to mammals [5, 10] (Figure 1).

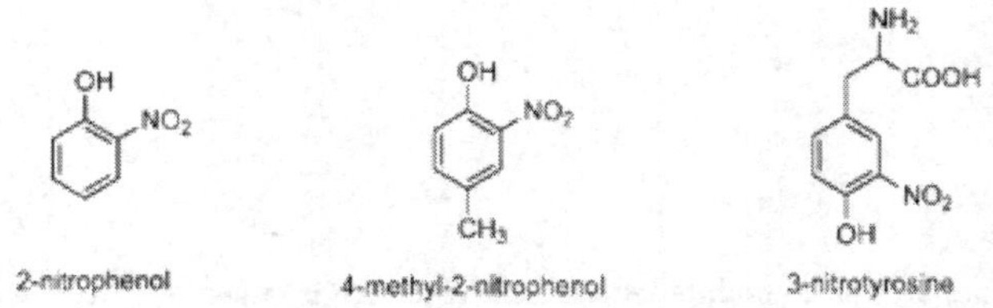

OH
NO_2
2-nitrophenol

OH
NO_2
CH_3
4-methyl-2-nitrophenol

NH_2
COOH
NO_2
OH
3-nitrotyrosine

Figure 1. Nitroaromatics as cell signaling molecules.

In nature, several nitroaromatic compounds are produced by the incomplete combustion of fossil fuel, which releases hydrocarbons. These hydrocarbons produce nitroaromatic compounds after nitration with nitrogen dioxide present in the atmosphere. Mixtures of nitropolyaromatic hydrocarbons can be produced to form 3-nitrobiphenyl,1- and 2-nitronaphthalene,3-NT and nitrobenzene by a hydroxyl radical-initiated mechanism [11–15].

1.2. Synthetic nitroaromatic compounds (production and uses)

The versatile chemistry of the nitro group ensures that nitroaromatic compounds serve as important feed stocks in different industrial processes. These compounds are commonly used in the manufacture of pharmaceuticals. For example, substituted nitrobenzenes and nitropyridines are used in the production of indoles, which are active components of several drugs and agrochemicals [16]. Paracetamol (an analgesic and antipyretic) is synthesized in a one-step reductive acetamidation from 4-nitrophenol [17]. Nitrobenzenes or halonitrobenzenes are used in the synthesis of derivatives of phenothiazines, a large class of drugs with antipsychotic properties [18, 19].

Some nitroaromatics like nitrobenzene, nitrotoluene, and nitrophenols are used in the synthesis of pesticides. For example, fluorodifen [20], bifenox, parathion [21], and carbofuran [22] are synthesized from nitrophenols. Some dinitrophenols like 2,5-dinitro-*o*-cresol have been used in the synthesis of herbicides, insecticides, fungicides, etc. [5, 23].

Aromatic amines are the largest feedstock group for chemical industries. It is estimated that the worldwide consumption of aniline is approximately 3 million tons [5]. This consumption grew by 7% annually till 2014 and is expected to reach 6.2 million tons in 2015 (Global Analysts report 2014 on aniline production). Aniline is used in the synthesis of drugs, pesticides, and explosives and used as a building block for the production of polyurethane foams, rubber, azo dyes, photographic chemicals, and varnishes [24].

Some nitroaromatics are used in the production of explosives like trinitrotoluene (TNT), which is produced by sequential nitration of toluenes. 1,3,5-Trinitrophenol (picric acid), which was prepared in 1771 as a yellow dye for fabrics [25], has also been used in explosive shells. The methyl group of TNT can be eliminated to produce trinitrobenzene (TNB), a high-energy explosive with decreased shock sensitivity [26].

1.3. Release of nitroaromatics in the environment

The estimated annual production of nitroaromatic compounds is 10^8 tons (http:www.ucl/agro/abi/gebi). Chemical industries release these compounds into the environment through various sources, like the use of pesticides and the improper handling or storage of chemicals. The leakage into industrial effluents by improper disposal or through accidental spills by explosive ammunitions are commonly responsible for these compounds to find their way into the environment [27]. A recent example of an industrial accident is from China, where about 100 tons of benzene and nitrobenzene were released to Songhua River because of an accidental explosion in the factory of China National Petroleum, Jilin City, on November 13, 2005 [28].

1.4. Toxic effects

The electron withdrawing property of the nitro groups creates a charge on the molecule. It is a unique property that makes the nitro group an important functional group for different industrial synthetic processes. Simultaneously, the same property makes these molecules hazardous to the environment. This is why these compounds are given hazardous rating 3 (HR 3), where 3 shows the highest level of toxicity [29]. These are toxic to most living organisms, including humans, fishes, algae, and microorganisms [30, 31]. Their toxicity principally manifests itself due to their ability to uncouple photo or oxidative phosphorylation processes [32, 33]. Some of these compounds are also known for their ecotoxicity [34, 35], immunotoxicity [36], carcinogenicity [37], mutagenicity [38, 39], and teratogenicity [40, 41]. Some nitroaromatics are also converted into carcinogens and mutagens when metabolized by liver or intestinal microflora [42, 43].

1.5. Treatment options

1.5.1. Physical, chemical, and physicochemical methods

Different physical methods are available for the treatment of these toxic chemicals, like adsorption, incineration, photo-oxidation, hydrolysis, volatilization, etc. During adsorption, these compounds are only adsorbed on resins and separated but not destroyed completely. In incinerations, these compounds are treated at very high temperatures, which is neither cost effective nor eco-friendly because toxic NO_x fumes are often released in the environment in this process. There are various reports on advanced oxidation processes (AOPs) that utilize ozone, UV radiation, hydrogen peroxide, or combinations of all these for the treatment of nitroaromatic compounds [44–47]. The use of hydrogen peroxide in these processes generates toxic intermediates and is therefore not cost-effective [48]. Because of limitations of these methods, biodegradation has emerged as a viable alternative.

1.5.2. Bioremediation

Bioremediation involves biological agents to catalyze the degradation and transformation of recalcitrant molecules to simpler structures. Few common terms used in these processes are defined as follows:

- Biodegradation is the breakdown of organic pollutants due to microbial activity. In this process, the microbe feeds on the pollutant to grow. The degradation of contaminant generates energy and microbe utilizes this energy for its growth.
- Biomineralization is the process of complete biodegradation. The organic contaminants are degraded completely through a series of degradation steps and finally converted to inorganic molecules like H_2O and CO_2. In the process, organic molecule provides both carbon and energy to the microbe, and if organic molecule is nitroaromatic, it provides nitrogen as well.
- Biotransformation is the process where in one organic molecule is modified by the action of biological agents. Sometimes, biotransformation occurs with cometabolism, where a microbe uses a substrate for its growth but transforms another substrate, which is not utilized by microbe for its growth.

Microbes have been isolated from almost all the parts of biosphere. Further, their adoptability for different environmental conditions and ability to utilize even recalcitrant compounds for their foods make them suitable agents for bioremediation.

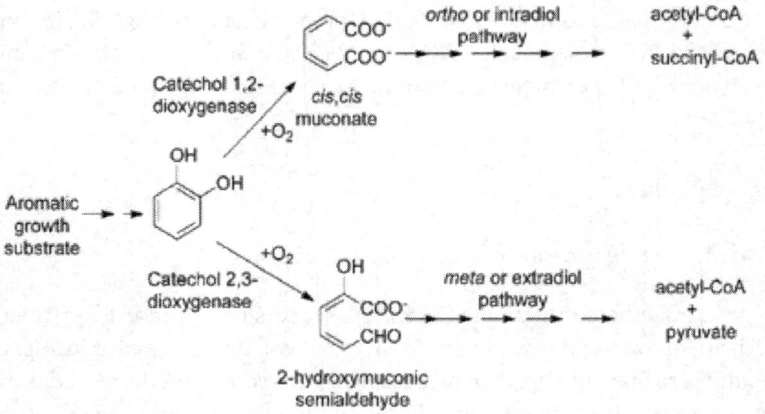

Figure 2. General principle of aerobic aromatic catabolism in bacteria. The three stages are as follows: the conversion of the growth substrate to catechol (or substituted catechol), then ring cleavage, and finally metabolism of the ring cleavage product to central metabolites by either the *ortho* or *meta* pathways. (Adapted from Williams and Sayers [51])

Biodegradation gained worldwide attention to treat toxic compounds [49]. This is because of its eco-friendliness. Supplementing the medium with readily utilizable carbon sources can enhance degradation processes. Thus, toxic intermediates are not generated and complete removal of toxic compounds is possible. During the last few decades, extensive research has been carried out for isolation of microbes with the abilities of degrading wide range of toxic nitroaromatic compounds and has been reviewed nicely [5, 49, 50]. Some common routes adopted by bacterial strains in nature during the degradation of aromatic compounds are described here.

1.6. General principles of pathways for aerobic aromatic catabolism

The pathway for catabolism of aromatic compound basically has three stages (Figure 2) [51–53]. In the first stage, the substrate undergo changes in its substituent groups by the action of mono- or dioxygenases to form catechols (or substituted catechols). The catechols then serve as substrates for the second stage of catabolism, that is, the ring opening. This process is facilitated by the action of dioxygenases, which breaks carbon–carbon bond by adding molecular oxygen and produce unsaturated aliphatic acid.

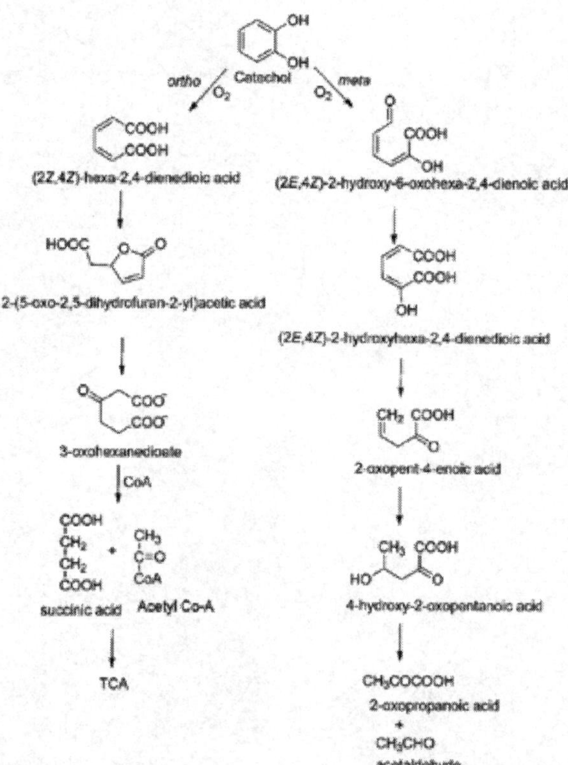

Figure 3. Central catechol pathway for the aerobic degradation of aromatics.

There are two families of ring opening enzymes, *ortho* or intradioldioxygenases, which produce *cis,cis* muconic acid (or its derivative), and *meta* or extradioldioxygenases, which produce 2-hydroxymuconic semialdehyde (or derivative). Both pathways (*ortho* and *meta* cleavage) are shown in Figure 3 and *meta* cleavage pathway for methylcatechols is shown in Figure 4. The third stage of catabolism is the conversion of the ring cleavage products into smaller compounds that can enter into central metabolic routes.

In general, aromatic compounds are initially catabolized by various pathways (known as peripheral pathways), which converge on a limited number of common intermediates (catechols or its derivatives). These intermediates are further utilized by a small number of common pathways (central pathways).

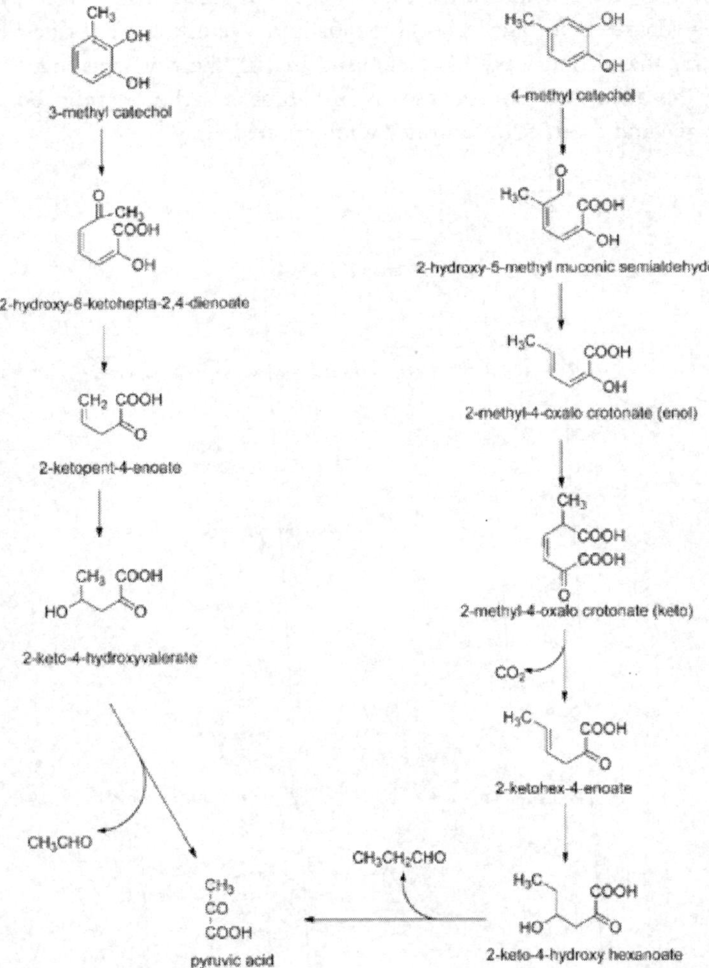

Figure 4. The *meta* cleavage pathway of methylcatechols.

2. Microbial degradation of nitroaromatics

When a nitroaromatic compound is exposed to the environment, its biodegradation takes place either by anaerobic route or by aerobic route. Different strategies applied for the degradation of nitroaromatic compounds by bacterial strains are shown in Figure 5.

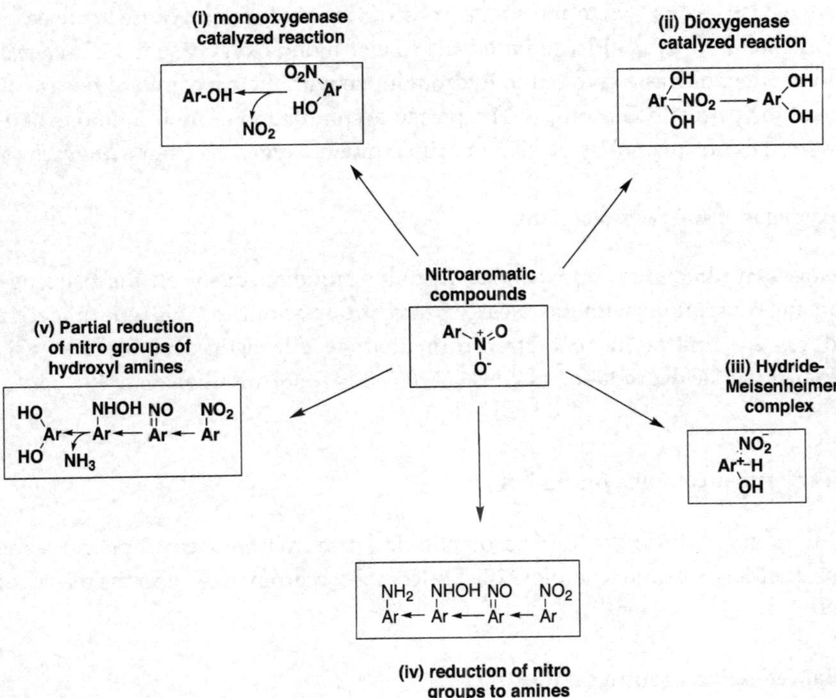

Figure 5. Different strategies of microbial remediation of nitroaromatics. (Adapted from Kulkarni and Chaudhari [49])

2.1. Anaerobic biodegradation of nitroaromatics

In this process, nitro group is reduced to nitroso derivative, hydroxyl amines, or amines by the action of nitroreductases. The degradation of most of the (poly)nitroaromatic compounds occurs only under anaerobic conditions [49, 54, 55]. The complete mineralization of nitroaromatic by a single anaerobic strain is very rare [56]. There are several reports showing that the initial step during the degradation of mono-, di-, and trinitroaromatic compounds is the reduction of nitro groups to amino groups [56–60].

2.2. Aerobic biodegradation of nitroaromatic compounds

Mono- and dinitroaromatics are mainly subjected to aerobic biodegradation and achieve to complete mineralization. Here nitroaromatics serve as source of carbon, nitrogen, and energy for the microbe. During the past few decades, several reports came up with isolation of microbes mineralizing different nitroaromatic compound and their degradation pathway. Few of them are extensively studied and characterized. There are different strategies in the aerobic degradation of nitroaromatics [61], which is used in nature as shown in Figure 5.

2.3. Reactions catalyzed by mono-oxygenases

Mono-oxygenases are known to add single oxygen atom at a time and cause the release of nitro group. Simpson and Evans [62] reported the role of mono-oxygenase in a *Pseudomonas* sp., where 4-nitrophenol was converted to hydroquinone with the concomitant release of nitrite. Subsequently, Spain and Gibson (1991) reported accumulation of hydroquinone and release of nitrite from 4-nitrophenol by partially purified mono-oxygenase from a *Moraxella* sp.

2.4. Dioxygenase catalyzed reactions

Dioxygenases are known to add two -OH groups simultaneously on the benzene ring of nitroaromatic compounds with the release of nitro group as nitrite. This type of mechanism is reported for 2,6-dinitrotoluene biotransformation by Alcaligenes eutrophus [63]. Other examples include the degradation of 2-NT [64, 65], 3-NT [66], nitrobenzene [67], and 2,4-DNT [68].

2.5. Meisenheimer complex formation

The addition of a hydride ion to the aromatic ring of nitroaromatic compound leads to the formation of a Meisenheimer complex [27]. The complex rearomatizes after the release of nitrite anion [69].

2.6. Partial reduction of nitro groups

In this mechanism, the nitro group is partially reduced to the corresponding hydroxylamine, which upon hydrolysis yields ammonia. This mechanism was reported in Comamonas acidovorans, where 4-nitrobenzoate is converted to 4-hydroxyl-aminobenzoate [69].

2.7. Different bacterial pathways reported for the degradation of mononitrotoluenes and nitrobenzene

2.7.1. Bacterial degradation of nitrobenzene

The aerobic degradation of nitrobenzene involves two major pathways: a most commonly found partial reductive pathway and a dioxygenase catalyzed pathway (Figure 6). In the oxidative degradation of NB, degradation starts with the action of nitrobenzene-1,2-dioxygenase, which converts nitrobenzene into catechol. This catechol is further cleaved by the action of catechol 2,3-dioxygenase and degraded by the *meta* cleavage pathway. This type of pathway is reported in *Comamonas* JS 765, *Acidovorax* sp. JS42, and *Micrococcus* sp. strain SMN1.

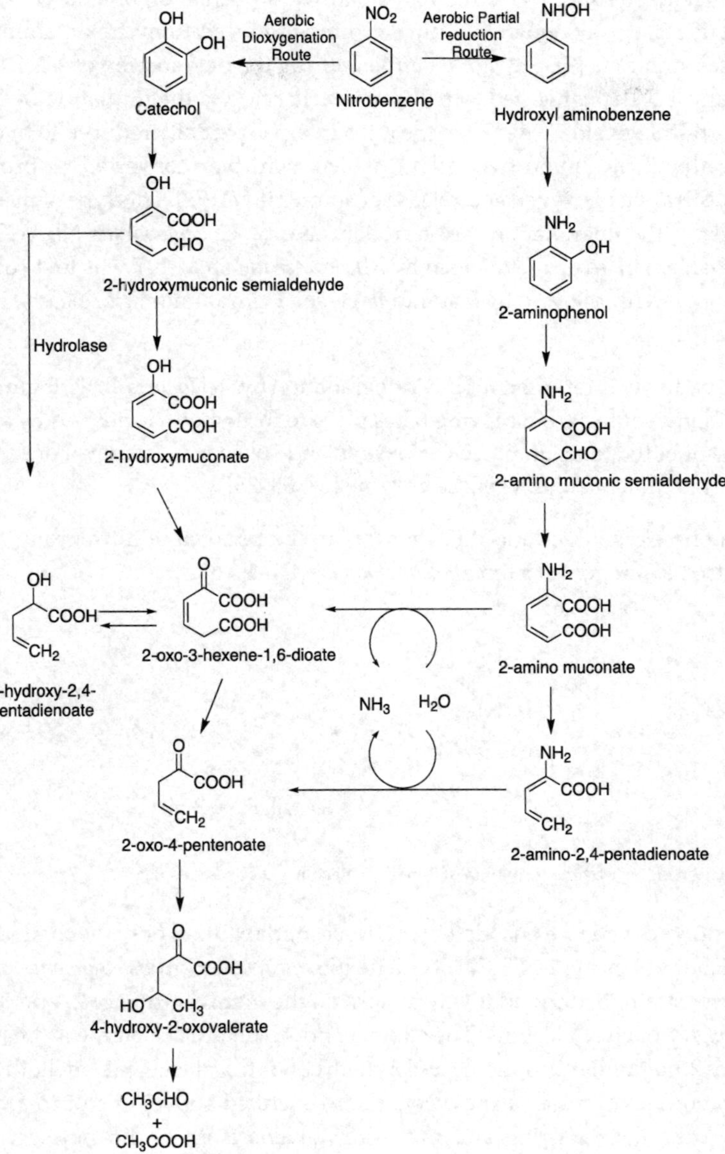

Figure 6. Degradation pathways for nitrobenzene. Aerobic dioxygenation route is reported in *Comamonas* JS765 [67]. The aerobic partial reductive pathway is from *Pseudomonas pseudoalkaligenes* JS45 [61].

2.7.2. *Bacterial degradation of mononitrotoluenes*

Different bacterial strains have been isolated from various sources, which can utilize nitrotoluenes as carbon source or both carbon and nitrogen source. There are several reports on different degradation pathways for mononitrotoluenes as described here.

Nitrotoluenes may be subjected to reductive pathways (formation of aminotoluenes) [70] or partial reductive pathway, wherein a nitro group is reduced to hydroxyl amino group and finally releases ammonia [71–72]. For example, during the degradation of 4-NT by *Pseudomonas* sp., initially 4-NT is converted into 4-nitrobezoic acid via the formation of 4-nitrobenzyl alcohol and 4-nitrobenzaldehyde. Then the nitro group is partially reduced to hydroxylamino derivative (rather than amino derivative), which is further converted to protocatechuate without the utilization of oxygen and release of ammonia [71, 72]. This type of mechanism was first reported for the degradation of 4-nitrobenzoate by *C. acidovorans* NBA-10 [69]. In yet another mechanism of 4-NT metabolism by *Micobacterium* sp., 4-NT was first converted to 4-hydoxyl aminotoluene followed by 6-amino-*m*-cresol. Here ammonia is released only after the ring cleavage [72].

An oxidative pathway is reported for 2-NT degradation by *Acidovorax* JS42 (Figure 7), wherein the initial oxidation of the aromatic ring takes place to form methylcatechols by simultaneous incorporation of both atoms of molecular oxygen and subsequent removal of the nitro group as nitrite by the action of a dioxygenase enzyme [64, 65, 73].

The role of mono-oxygenases and dioxygenases in the removal of nitro group from *p*-nitrophenol has also been reported from a *Pseudomonas* sp. [74–76].

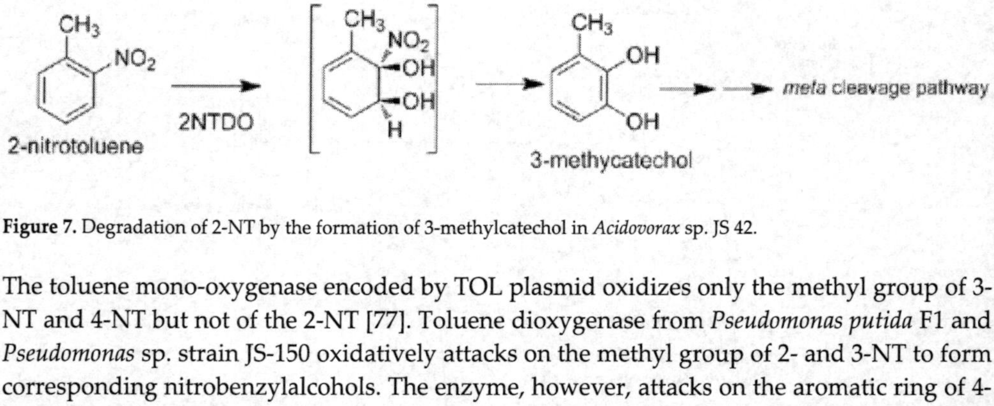

Figure 7. Degradation of 2-NT by the formation of 3-methylcatechol in *Acidovorax* sp. JS 42.

The toluene mono-oxygenase encoded by TOL plasmid oxidizes only the methyl group of 3-NT and 4-NT but not of the 2-NT [77]. Toluene dioxygenase from *Pseudomonas putida* F1 and *Pseudomonas* sp. strain JS-150 oxidatively attacks on the methyl group of 2- and 3-NT to form corresponding nitrobenzylalcohols. The enzyme, however, attacks on the aromatic ring of 4-NT to produce 2-methyl-5-nitrophenol and 3-methyl-6-nitrocatechol [81]. In both cases (either with toluene mono-oxygenase or dioxygenase as described above), the nitro group was not removed from the benzene ring and mononitrotoluene isomers did not serve as growth substrate. Degradations of monosubstituted 2-, 3-, and 4-nitrotoluenes were also reported from an adapted activated sludge system [79].

The two strains of *Comamonas* JS47 and JS46 capable of degrading 4-nitrobenzoate and 3-nitrobenzoate respectively were immobilized on alginate beads jointly and separately, and these beads were loaded in the reactor and fed to different regimes of alternating nitrobenzoate isomer or mixed nitrobenzoate isomer. Through this experiment, it was deduced that same beads containing both strains were able to recover faster from change in input composition than different beads containing different strains [80].

2.7.2.1. *Bacterial degradation of 4-nitrotoluene*

Two different pathways are reported in bacterial strains (as shown in Figure 8). In aerobic methyl group oxidation (*Pseudomonas* sp. strain TW3), initially 4-NT is converted to 4-nitrobenzoic acid, and then the nitro group is partially reduced to hydroxylamine derivative, which is further converted to protocatechuate without the utilization of oxygen but with concomitant release of ammonia [81]. In another mechanism, 4-NT degradation is followed by the formation of 4-hydroxylaminotoluene.

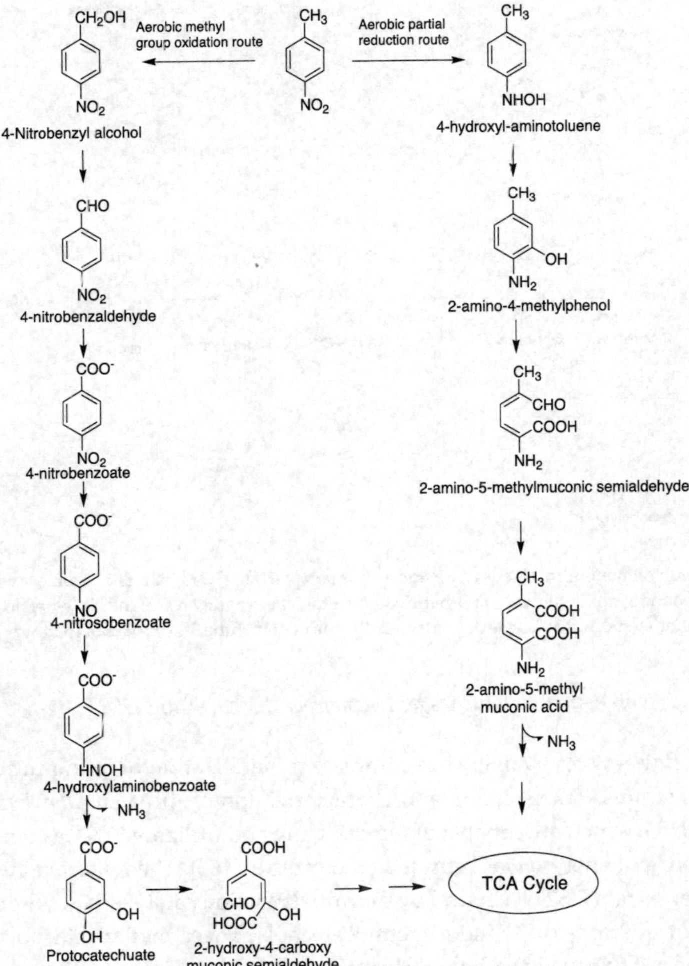

Figure 8. Degradation pathways for 4-nitrotoluene. Aerobic partial reduction rout is present in *Mycobacterium* sp. strain HL-4NT-1. Aerobic methyl group oxidation rout is reported in *Pseudomonas* strains TW3 and strain 4NT.

There are few well-characterized bacterial strains that degrades or biotransforms more than one mono-nitro compounds. 2-NT and 4-NT could transform to their corresponding aminotoluenes and hydroxyl nitrotoluenes (pathways A and B, Figure 9) in *P. putida* OU83. Both oxidative as well as reductive attack is reported during metabolism of 3-NT [71]. Here 70% of the 3-NT was reduced to aminotoluene, whereas 30% was converted to 3-nitrophenol via the formation of 3-nitrobenzylalcohol, 3-nitrobenzaldehyde, and 3-nitrobenzoic acid. 3-Nitrophenol was further metabolized with the release of nitrite (pathway C, Figure 9).

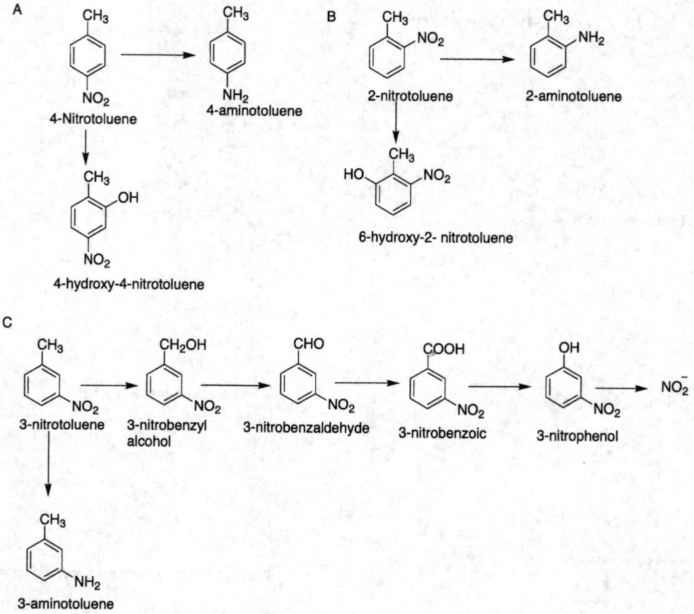

A
CH_3 NO_2 4-Nitrotoluene → CH_3 NH_2 4-aminotoluene
↓
CH_3 OH NO_2 4-hydroxy-4-nitrotoluene

B
CH_3 NO_2 2-nitrotoluene → CH_3 NH_2 2-aminotoluene
↓
CH_3 HO NO_2 6-hydroxy-2- nitrotoluene

C
CH_3 NO_2 3-nitrotoluene → CH_2OH NO_2 3-nitrobenzyl alcohol → CHO NO_2 3-nitrobenzaldehyde → COOH NO_2 3-nitrobenzoic → OH NO_2 3-nitrophenol → NO_2^-
↓
CH_3 NH_2 3-aminotoluene

Figure 9. Degradation of mononitrotoluenes by *Pseudomonas putida* strain OU83 [82]. The strain converts 2- and 4-nitrotoluenes to corresponding aminotoluenes and hydroxyl nitrotoluenes (pathways A and B), whereas 70% of 3-NT was converted to 3-aminotoluene and 30% was degraded via the formation of the 3-nitrophenol (pathway C).

2.7.2.2. Degradation of mononitrotoluenes by Diaphorobacter sp. strain DS2

There are very few reports available on complete mineralization of mono-nitroaromatics by single bacterial strains. The isolation and characterization of three *Diaphorobacter* sp. strains DS1, DS2, and DS3, which are capable of mineralizing and utilizing 3-NT as the sole source of carbon, nitrogen, and energy, was reported by our group [66]. The mineralization of 3-NT by *Diaphorobacter* sp. strain DS2 was found by the initial reaction catalyzed by a dioxygenase with the formation of mixtures of 3- and 4-methylcatechols. These methylcatechols were further degraded by the *meta* cleavage pathway. This strain was able to degrade other compounds like 2-NT, 4-NT, nitrobenzene (NB), 2CNB, and 3CNB through an oxidative degradation route. Cloning and sequencing of first enzyme of the pathways showed the presence of a multicomponent dioxygenase, i.e., 3NTDO. This 3NTDO gene was found to be present on the genomic DNA of the strain on a 5-kb DNA stretch. Its subunits were identified as a reductase, a ferredoxin, an oxygenase large and small subunit, and a regulatory gene product [83]. Subunits of 3NTDO were individually expressed in *E. coli* and purified by various purification techniques. To get active enzyme, all the subunits were mixed together in a certain proportion with added NADH. Its active recombinant-reconstituted enzyme showed the conversion of nitrotoluenes and nitrobenzenes to methylcatechols and catechols as measure products with the release of nitrite [84]. Some other minor products are also formed as shown in Table 1. Products other than (methyl)catechols are dead-end products and observed in the culture broth during the degradation of its corresponding substrates with the *Diaphorobacter* sp. strain DS2. This strain did not grow on any dinitrotoluenes (2,4 or 2,6), but recombinant-reconstituted 3NTDO released nitrite from 2,6-dinitrotoluene.

Substrate	Products formed by 3NTDO		
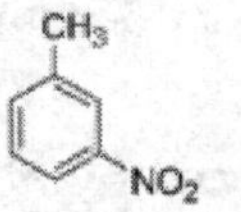 3-nitrotoluene	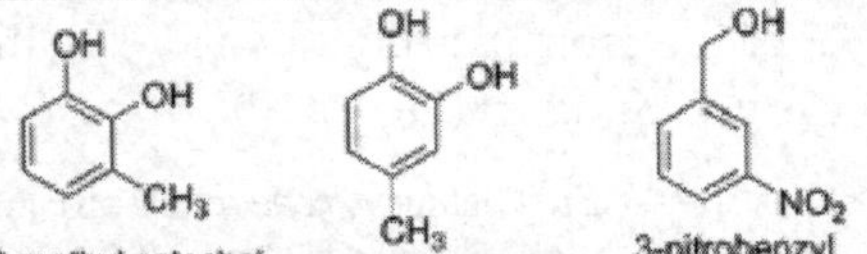 3-methyl catechol (54%)	4-methyl catechol (39%)	3-nitrobenzyl alcohol (7%)
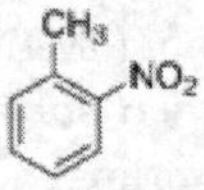 2-nitrotoluene	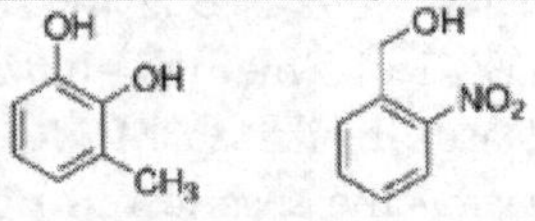 3-methyl catechol (91%)	2-nitrobenzyl alcohol (9%)	
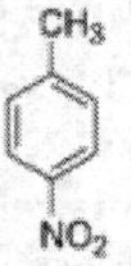 4-nitrotoluene	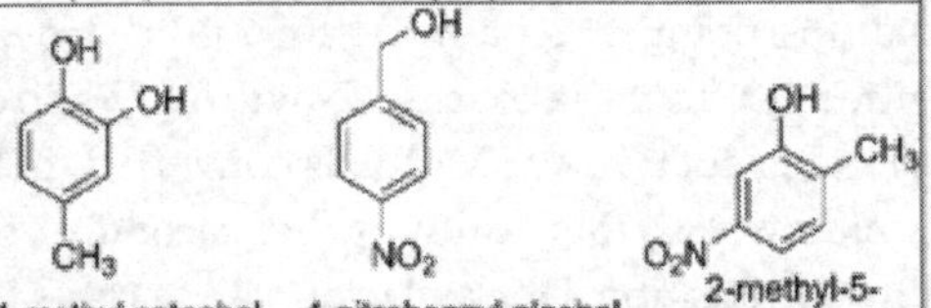 4-methyl catechol (34%)	4-nitrobenzyl alcohol (3%)	2-methyl-5-nitrophenol (63%)
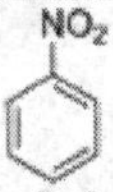 nitrobenzene	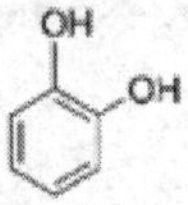 catechol		
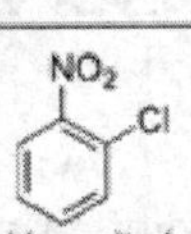 2-chloro nitrobenzene	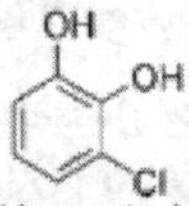 3-chloro catechol		
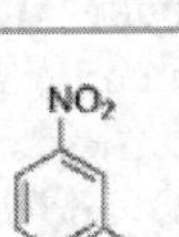 3-chloro nitrobenzene	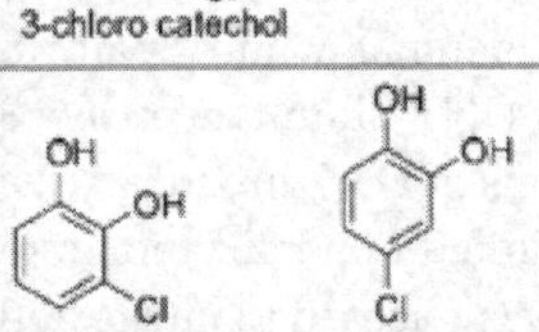 3-chloro catechol (8%)	4-chloro catechol (92%)	

Table 1. Products from the various substrates by recombinant 3NTDO.

2.8. Degradation of dinitrotoluene

2.8.1. 2,4- and 2,6-dinitrotoluene

Burkholderia sp. strain DNT and strain R34 mineralized 2,4-DNT with a nitrite removal pathway involving dioxygenase and mono-oxygenase enzyme. Here both nitro groups are not removed simultaneously but in a stepwise fashion. First, dinitrotoluene dioxygenase (DNTDO) attacks on the benzene ring converting DNT into 4-methyl-5-nitrocatachol (4M5NC) with the simultaneous removal of a nitrite group. Further, MNC mono-oxygenase removes another nitrite and converts the substrate into 2-hydroxy-5-methylquinone. These dioxygenase and mono-oxygenase enzymes have been cloned and characterized [68, 85]. This 2-hydroxy-5-methylquinone eventually leads to the formation of 2,4,5-trihydroxytoluene, a reaction that is catalyzed by HMQ reductase. Ring fission of 2,4,5-trihydroxytoluene likely occurs at position 5,6 of the aromatic ring to yield 2,4-dihydroxy-5-methyl-6-oxo-hexa-2,4-dienoic acid as ring cleavage product (Figure 10).

Nishino et al. [61, 86] isolated *Burkholderia cepacia* strain JS850 and *Hydrogenophaga paleronii* strain JS863 that were able to mineralize 2,4-DNT in the same way but degraded 2,6-DNT in a different way. When 2,4- and 2,6-DNT were used as the sole source of carbon and nitrogen together, the dioxygenation of 2,6-DNT to 3-methyl-4-nitrocatachol (3M4NC) was the initial reaction, accompanied by the release of nitrite. 3M4NC was then subjected to *meta*-ring cleavage (Figure 11) without releasing the second nitro group prior to the ring cleavage. Although 2,4-DNT-degrading strains also could convert 2,6-DNT to 3M4NC, further catabolism was halted at the point. The pathway for 2,4-DNT degradation was different from that for 2,6-DNT degradation. In the latter case, 2-hydroxy-5-nitro-6-oxohepta-2,4-dienoic acid was the first ring fission product. How 3M4NC is converted to 2-hydroxy-5-nitro-6-oxohepta-2,4-dienoic acid is unknown. In this pathway, the second nitrite group is released at the latter stages of the pathway. Position 3 methyl group appears as the determinant recognized by the initial dioxygenase to produce highly specific 3M4NC in the 2,6-DNT pathway proposed. The gene encoding this dioxygenase showed a nucleotide sequence similar to the α subunit among nitroarene dioxygenase.

The genes for the initial dioxygenases involved in 2,4-DNT and 2,6-DNT degradation are all closely related, but the enzymes are produced at low constitutive levels [27, 86]. After initial dioxygenation, the two pathways appear to diverge (Figures 10 and 11).

How does DNT degradation get affected by the presence of both isomers is important since 2,4-DNT and 2,6-DNT are produced in a 4:1 ratio [87]. These are often present together in munitions plant wastewater. Lendenmann and Spain [88] initially failed to observe the degradation of 2,4-DNT and 2,6-DNT simultaneously. Subsequently, an aerobic biofilm, initially fed with low concentrations of DNT mixture, was tested. These concentrations were then gradually increased and exhibited mineralization rates of 98% and 94% for 2,4- and 2,6-DNT, respectively. The nitrogen was released as nitrite, reflecting oxidative bacterial activity. Isomer concentration needed to be kept below inhibitory levels as high concentrations of each isomer inhibited the degradation of the other. The simultaneous degradation of 2,4- and 2,6-DNT may be unpredictable until an adapted population is established [87].

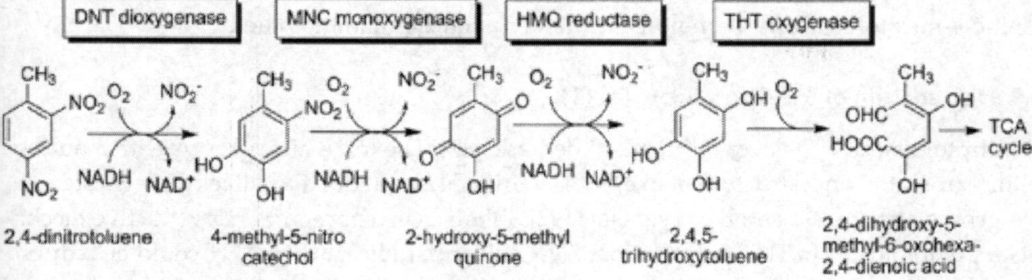

Figure 10. 2,4-DNT metabolism pathway in *Burkholderia* sp. strain DNT and strain R34.

Although nitrotoluene degraders are widely distributed at contaminated sites, the contaminants still persist for very long periods, leaving unanswered questions as to why biodegradation is ineffective to remove them. Efficient anaerobic pathways for the degradation either of mono- or dinitrotoluenes are not known, and 2,3-DNT currently does not appear to be degradable [87].

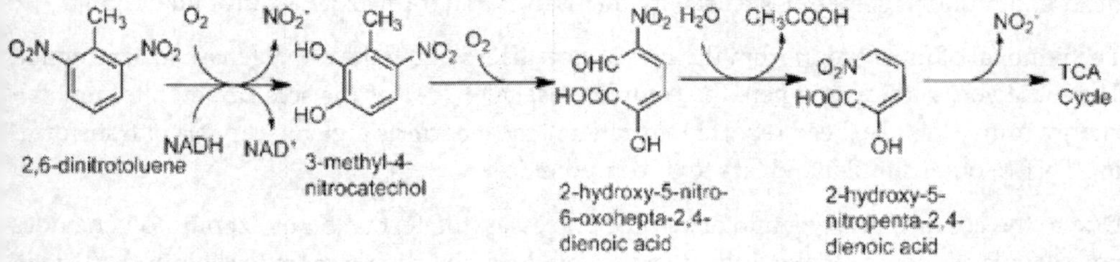

2,6-dinitrotoluene
3-methyl-4-nitrocatechol
2-hydroxy-5-nitro-6-oxohepta-2,4-dienoic acid
2-hydroxy-5-nitropenta-2,4-dienoic acid
TCA Cycle

Figure 11. 2,6-DNT metabolism pathway in *Burkholderiacepacia* strain JS850 and *Hydrogenophaga paleroni* strain JS863.

Bae et al. [89] found that in an anaerobic fluidized-bed granular carbon bioreactor, 2,4 DNT can be converted completely to 2,4-diamino toluene, which subsequently mineralized in batch activated sludge reactor. Paca et al. [90] took a mixture of microbes found in the mononitrotoluene, 2,4-dintrotoluene, and 2,6-dinitrotoluene contaminated soil. These microbes were extracted and immobilized on the packing material of the packed bed reactor (PBR). Varying concentrations of 2,4-DNT and 2,6-DNT were used. In this case, two types of packing material were used out of which the reactor packed with Poraver removed 97% DNT in 11 days and the one packed with fine clay achieved the efficiency of 78%. After 20 days, the metabolites detected were 2-amino-4-nitrotoluene and 2,4-diamino toluene.

Wang et al. [91] reported that in wastewater enriched with contaminated DNT taken from Qingyang chemical industry with a DNT concentration of 3.55–95.65 mg/L, ethanol was mixed in the wastewater to act as an electron donor. The reactor in this case was made of polymethyl metacrylate containing polyurethane foams for microorganism immobilization. The microorganism used was B925. Initially, the reactors were domesticated and immobilized with microorganism for first 10 days, and then the whole system was further operated for 140 more days, gradually increasing the concentration of 2,4 DNT. DNT was then transformed to 2-amino-4-nitrotoluene and 4-amino-2 nitrotoluene and 2,4-diaminotoluene.

2.9. Degradation of Trinitrotoluene (TNT)

Trinitrotoluene (TNT) is very difficult to degrade [87]. The three nitro groups with a nucleophilic aromatic ring structure make TNT vulnerable to reductive attack but resistant to oxygenase attack from aerobic organisms [92]. In most current reports, the reductive mechanism predominates in TNT degradation. New evidence indicates that TNT could be reduced by carbon monoxide dehydrogenase from *Clostridium thermoaceticum* [93] and by the manganese-dependent peroxidase (MnP) from the white-rot fungus *Phlebia radiata* [94]. Based on the discovery of pentaerythritoltetranitrate (PETN) reductase from *Enterobacter cloacae* PB2, French et al. [95] found that this strain could grow slowly on 2,4,6-TNT under aerobic conditions as the sole nitrogen source without the production of dinitrotoluene as an intermediate and catalyzed conversion of the TNT via a hydride–Meisenheimer complex with the nitro group released as nitrite.

Bacteria basically tends to biotransform TNT to aminonitroaromatic compounds through aerobic degradation, which in many cases turn out to be the dead-end products, and this reduced dead-end products sometimes react with themselves and form azotetranitrotoluene [96].

The removal of nitro group from the ring is essential to allow the dioxygenase to act upon it. There are very rare cases where the complete usage of TNT as the sole carbon nitrogen and energy source has also been reported. In general, most bacteria are only capable of transforming TNT to other simpler and less toxic compounds.

Due to the absence of oxygen in anaerobic processes, the formation of azonitrotoluene does not take place, thereby making the degradation through bacteria more feasible and efficient. Reduction product of TNT is very prominent in case of anaerobic process, which easily forms triaminotoluene, which is far less toxic and more soluble in water than TNT (Figures 12 and 13).

Collie et al. [116] observed the biodegradation of TNT in the liquid phase bioreactor by four different bacterial strain having pure TNT in a liquid medium. The initial concentration of TNT (70mg/L) was periodically extracted from the bioreactor for by-product identification with the help of HPLC. The bacteria used were one strain of *Enterobacter*, one strain of *Pseudomonad*, and two strains of *Alcaligenes*. Two basic intermediates, i.e., 2-amino-4,6-DNT and 4-amino-2,6 DNT, were observed with all the bacteria after 12 hours.

A bench-scale reactor by Cho and coworkers [114] used *P. putida* HK6 (collected from RDX-contaminated soils). This was used for the degradation of several nitroaromatic compounds simultaneously in one go, i.e., TNT–RDX–atrazine–simazine (TRAS). Cells were grown in a liquid media composed of 30–100 mg TNT, 5–15 mg RDX, 20–50 mg atrazine, and 5–15 mg simazine and other basal salts like K_2HPO_4, NaH_2PO_4, $MgSO_4\,7H_2O$, $CaCl_2\,2H_2O$, $FeCl_3\,6H_2O$ in appropriate quantities, and subsequent experiments on the utilization of nitroaromatic compounds were carried out using a 2.5-L bottom driving type bench-scale reactor with a water condenser at 5°C and operated at 30°C at 150 rpm. A 10% inocula of test culture was

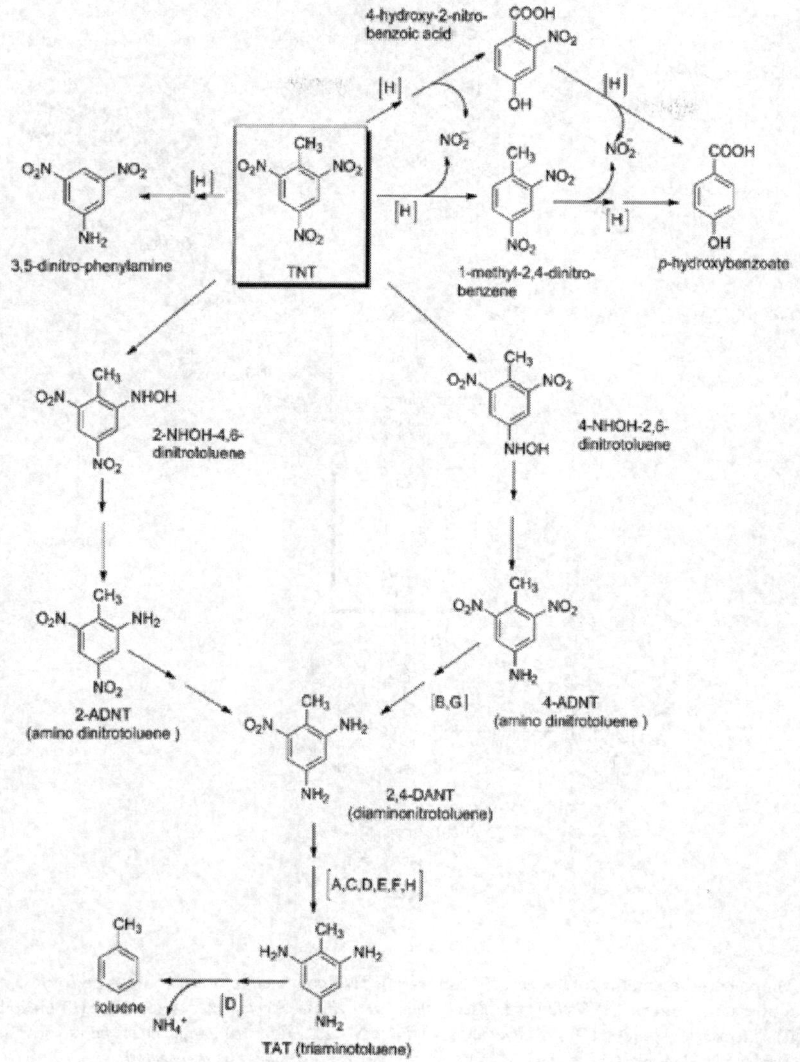

Figure 12. Different reductive pathways in different bacteria: (*A*) *Clostridium acetobutylicum, Escherichia coli, Lactobacillus* sp. [97]; (*B*) *Clostridium bifermentans* CYS-1 [98]; (*C*) *Clostridium bifermentans* LJP-1 [99]; (*D*) *Disulphovibriosp* strain B [100]; (*E*) *Disulphovibrio* sp. [101]; (*F*) *Disulphovibrio* sp. [102]; (*G*) *Methanococcus* sp. strain B [103]; (*H*) *Veillonella alkalescens* [57].

grown with nitroaromatic compound, and the turbidity observed showed that the bacteria *P. putida* HK6 was able to degrade 100 mg/l TNT, 15 mg/l RDX, 50 mg/l atrazine, and simazine in 4, 24, 2, and 4 days after incubation, respectively. In this experiment, it is noteworthy that the presence of Tween-80 in the culture led to the complete degradation of TRAS compounds, whose otherwise partial degradation was TNT (80%), RDX (35%), and simazine (78%) during the incubation period [114].

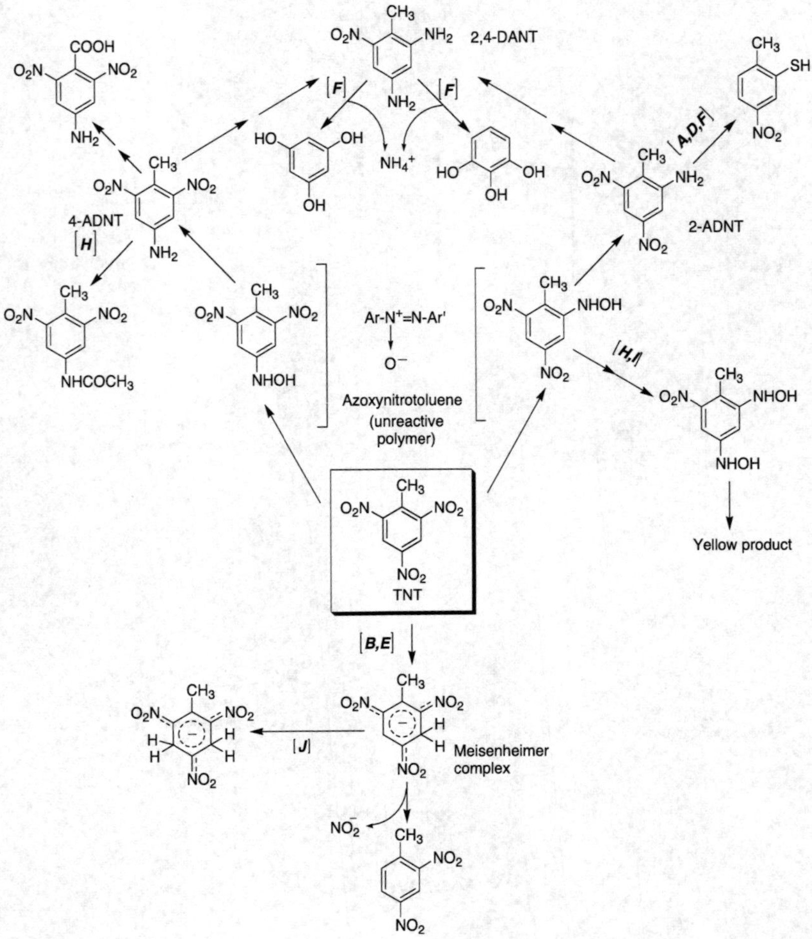

Figure 13. Complete representation of the aerobic pathway of TNT conversion to its products performed by the group of bacteria shown in the figure: (*A*) *Bacillus* sp., *Pseudomonas aeruginosa, Staphylococcus* sp. [104]; (*B*) *Enterobacter cloacae* PB2 [105]; (*C*) *Enterobacter* sp. [89]; (*D*) *Pseudomonas aeruginosa* MA101 [106]; (*E*) *Pseudomonas florescence* [107]; (*F*) *Pseudomonas florescence* B3468 [108]; (*G*) *Pseudomonas florescence* [109]; (*H*) *Pseudomonas pseudoalcaligenes* [110]; (*I*) *Rhodococcus erythropolis* [111]; (*J*) *Serratia marcensens* [112].

Similarly, the degradation of some other nitroaromatic compounds was reported on the basis of field trials. A highly active microbial consortium was chosen by Oh et al. [115] for field trials to degrade the wastewater samples having 4,6-dinitro-*ortho*-cresol (DNOC) [116]. Fixed bed column reactors were employed to increase the volume density of active biomass and to degrade DNOC in wastewater stream. In this case, glass bead matrix was used to immobilize the bacterial colonies. The mixed bacterial culture CDNOC1-3 metabolizes DNOC with concomitant liberation of nitrate. The efficiency of the bioreactor was 86%. This was substantially higher over a batch culture mode (60–65%). The following schematic diagram (Figure 14) shows the fixed bed column reactor system [116].

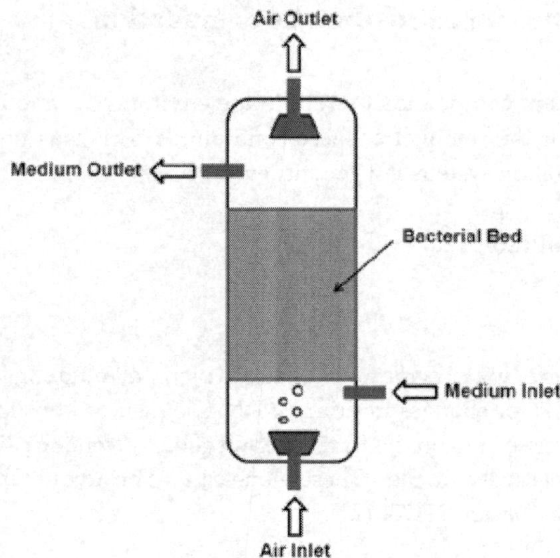

Figure 14. Schematic representation of the reactor used for the degradation of 4-nitrobenzoate and 4-aminobenzoate by *B. cepacia* PB4 in the abovementioned study [59].

A similar kind of experiment was performed in year 1999 in which *B. cepacia* strain PB4 was isolated from 4-aminobenzoate. This also possessed the capability to degrade 4-nitrobenzoate. Thus, considering both classes of the contamination to be toxic and mutagenic, an efficient strategy of decontamination was applied because *B. cepacia* was able to use both as the sole source of carbon, nitrogen, and energy. In order to prevent toxic effects, these compounds were supplied in lower concentration, i.e., 10–100 ppm and also in order to increase the efficiency of the procedure at such low concentration the degradative bacteria was immobilized on porous diatomaceous celite. This degradation was carried out in a packed bed reactor (PBR). The bioreactor consisted of a glass tube (296 × 41 mm) filled with 150 mm packed bed of celite-grade R-633 or R-635, which was placed at 30°C. It was shown that the nitroaromatic and aromatic amino compounds, which are otherwise unlikely to degrade together if present in any affected area, had been degraded simultaneously by single microorganism supply.

The eventual objective of all the biochemical and molecular characterization of the bacterial degradation of pollutants is to develop strains, which could be used in the bio remediation process. In this respect, another good field trial experiment was described by Labana et al. [117, 118] with bacterial strain *Arthrobacter protophormiae* RKJ100. The result clearly showed that the disappearance of the nitroaromatic pollutants.

Similarly, *Pseudomonas* sp. ST53 was also used as a microbe to degrade TNT and other explosives, but it is best suited on land and water only when the contamination is low [119]. Qureshi et al. [120] reported a bacterial strain *Arthrobacter* sp. HPC1223, which was capable of degrading 2,4,6-trinitrophenol prominently. This also poses the capability to degrade dinitrophenol and mononitrophenol showing broad substrate specificity.

3. Evolution of genes for nitroaromatic degradation

Nitroaromatics are recent compounds present in the environment and bacterial strains that adapted themselves for the removal of these compounds. This was possible only through evolution of its degradation system at a genetic level.

3.1. Modes of gene evolution

3.1.1. Mutational drift

Substrate profile of an existing enzyme may be altered by point mutations in its corresponding genes [124]. The reasons for changes in primary DNA sequences are slippage of DNA polymerase while replications, erroneous DNA repair, and gene conversion [122]. However, results of these changes are relatively smaller. These alone cannot be accountable for adaptation to the new environment by bacteria [123, 121].

3.1.2. Genetic rearrangement within a cell

The rearrangement of genes for the development of new pathway may take place by the help of cells own recombination system. Gene segments can be exchanged between two positions flanked by homologous sequences, insertion elements, transposons, and even sequence identities of four base pairs are sufficient to facilitate this process [122].

3.1.3. Horizontal gene transfer

Horizontal gene transfer is reported as the main source of evolution of pathways in bacteria [123, 124]. Sequencing results of genomes from different bacterial strains have revealed the presence of acquired genes in mosaic like fashion throughout bacterial genomes. Their presence varies from almost negligible (*Rickettsia prowazekii* and *Mycobacterium genitalium*) to about 17% (in *Synechocystis* strain) [123]. Plasmids also play a major role in carrying catabolic genes during horizontal gene transfer [125]. Transposons are also known to facilitate the catabolic gene transfer processes [126].

3.2. Nitroaromatic degradation pathway as a role model for study the gene evolution

A role model to study evolution of microbial pathways is to study the degradation of nitroaromatic compounds in different bacteria.

Nitroaromatic compounds are relatively new to the environment, but bacterial systems have already evolved the ability to metabolize them. This cannot be possible only by spontaneous, independent evolution of several new enzymes in a single bacterium. Horizontal gene transfer has to play a key role in combination with the mutagenesis of the existing enzymes to facilitate rapid evolution of new pathways. Evolution of diverse pathways for the degradation of different nitroaromatics thus stands testament to this.

A good example of this is the evolution of chloronitrobenzene dioxygenase system from a chloronitrobenzene degrading strain *Pseudomonas stutzeri* [127]. It has several insertion sequences embedded between the gene clusters, which proves such involvement in its evolution. In its dioxygenase enzyme system, reductase and ferredoxin seems to have come from different origins because its reductase and ferredoxin share maximum identity with anthranilate dioxygenase, which is a type IV oxygenase, whereas its oxygenase subunits show a maximum identity with nag (naphthalene degradation gene) and nitroarene dioxygenase, which falls under type III oxygenase systems. This enzyme system thus best illustrates the evolution of catabolic genes best because its upper pathway enzymes seem to have originated from a nitroarene degradation pathway and its lower pathway genes have evolved from some chloroaromatic compound degradation pathway. The genes responsible for these pathways are present in a patchwork like assembly in *P. stutzeri* [127]. The presence of insertion elements in the gene cluster confirms its role in the formation of a modular assembly and its role in evolution of the gene cluster.

Another example is the origin of 2,4-DNT degradation pathway in *B. cepacia* R34 where enzymes for the degradation pathway have originated from at least three different sources [128]. The first enzyme in the degradation pathway (DNTDO), which removes the first nitrite from 2,4-DNT, seems to have originated from naphthalene degradation pathway like in *Ralstonia* sp. strain U2. [129]. However, 4-methyl-5-nitrocatechol mono-oxygenase, which facilitates the removal of the second nitrite, appears to be derived from a pathway for degradation for chloroaromatic compounds [128]. The last enzyme of the pathway could have originated from a gene cluster for amino acid degradation. This is known as progressive compaction of the genes. However, the presence of ORFs without any known role in 2,4-DNT degradation and truncated transposons in the regions suggests that compaction is probably in an intermediate stage in the evolution of such an optimal system with the genetic materials from different bacterial origin.

A good example of similar type of evolution is reported in diphenylamine degrader *Burkholderia* sp. strain JS665 [130]. In this pathway, diphenylamine is converted to catechol and aniline. An analysis of this sequence of diphenylamine degrading enzyme system showed that it has evolved by recruiting two pathway enzymes, one of which is from dioxygenase and the other is from nitroaniline degradation pathway enzymes in a much more recent evolution event [127].

Another example of evolution of genes for nitrotoluenes degradation is the evolution of 3NTDO in *Diaphorobacter* sp. strain DS2. Five complete ORFs were identified by probable ORF finding program and by homology to polypeptide sequences from several previously reported multicomponent dioxygenase systems. The predicted translation products from ORFs were designated as a putative regulatory protein, a ferredoxin reductase subunit, a ferredoxin subunit, an oxygenase large, and a small subunit based on their homology. A 571-bp DNA stretch was present in between reductase and ferredoxin subunits. In its gene structure, the regulatory protein is divergently transcribed from the other four ORFs. The organization of gene cluster and its similarity with other known dioxygenases is shown in Figure 15.

It has been suggested in several reports that *Ralstonia* sp. strain U2 [129] is the progenitor of all the nitroarene dioxygenases because it has the entire functional gene in its gene assembly.

This seems true for the 3NTDO as well because the sequence present in between the reductase and the ferredoxin is truncated into parts of two functional ORFs present in the *Ralstonia* sp. strain U2. The regulatory protein sequence of 3NTDO (MntR) differs only at three amino acid positions from NagR of *Ralstonia* sp. strain U2 out of which two are uniquely present in strain DS2 only.

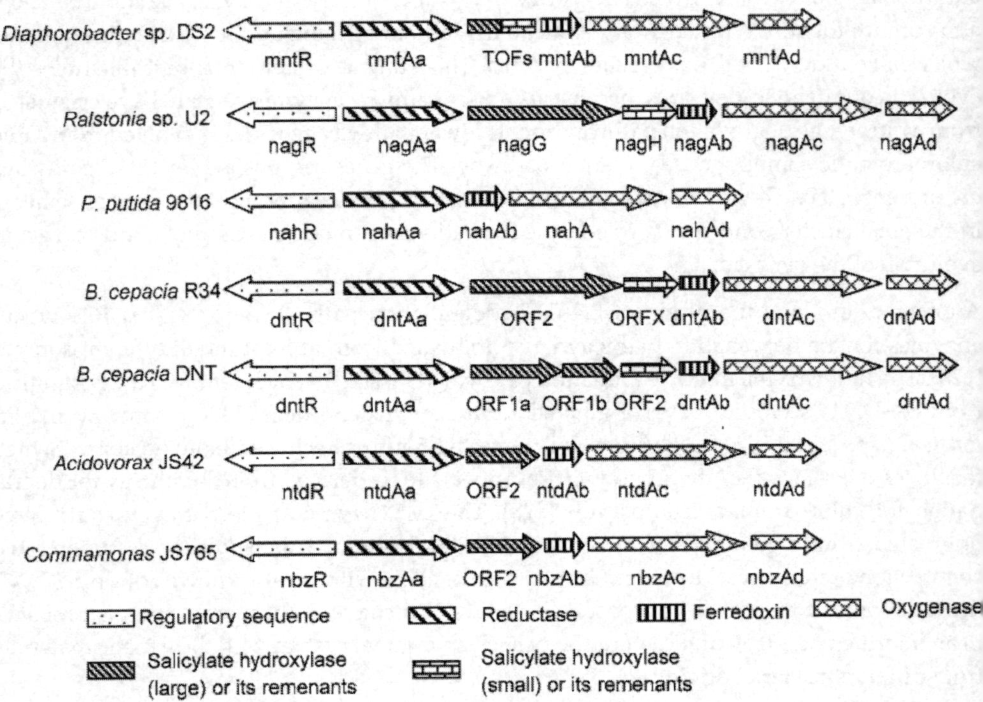

Figure 15. Comparison of gene arrangements in various multicomponent dioxygenases.

The different components (reductase, ferredoxin, and oxygenase) of 3NTDO show different levels of sequence identity with components from similar multicomponent enzyme systems of different organisms. Its reductase subunit (MntAa) shares a high amino acid sequence identity with those of DNTDO from *B. cepacia* [128] and NDO of *Ralstonia* sp. strain U2 (99%) [129], but its ferredoxin subunit (MntAb) is 100% identical to the ferredoxin of dinitrotoluene dioxygenase from *Burkholderia* sp. strain DNT and *Burkholderia* sp. strain R34. Its large oxygenase subunit (MntAc) showed more identity with chloronitrobenzene dioxygenase (CnbAc, 96%) from *P. stutzeri* ZWLR2-1, whereas the small oxygenase subunit (MntAd) showed more identity with PahAd of *Comamonas testosteroni* (96%) and NTDO from *Acidovorax* sp. strain JS42. It is known that oxygenase large subunit controls substrate specificity. If we compare important active site residues in oxygenase large subunit (MntAc) of strain DS2 with well-characterized oxygenase systems, it contains amino acid combinations of other systems in which the sequence retains His293, which is present in NDO system of *P. putida* 9816-4, *Ralstonia* U2, *C. testosteroni* H, and *Burkholderia* sp. C3. Position 350 is occupied by Valine, which is reported in the DNTDO of *Burkholderia* sp. strain R34. Thus, the above facts seem to indicate that 3NTDO gene in *Diaphorobacter* sp. strain DS2 came through a horizontal gene transfer from ancestors common to strains like *Ralstonia* U2 or *Burkholderia* sp. strain R34, and then its catalytic subunit has been diversely evolved to degrade other nitroaromatic compounds.

The mechanism by which enzymes for the degradation of synthetic compounds have evolved so rapidly still cannot be explained only by horizontal gene transfer and mutations. It can be explained in part by the term promiscuity. Promiscuity refers to the ability of a protein to perform dual functions using same active site [131, 132]. Protein evolution toward a new function based on promiscuity involves transition of an existing specialized enzyme to a generalized intermediate enzyme and then into a new specialized enzyme (Figure 16). A good example of this is transcriptional regulator found in nitroarene dioxygenases [133].

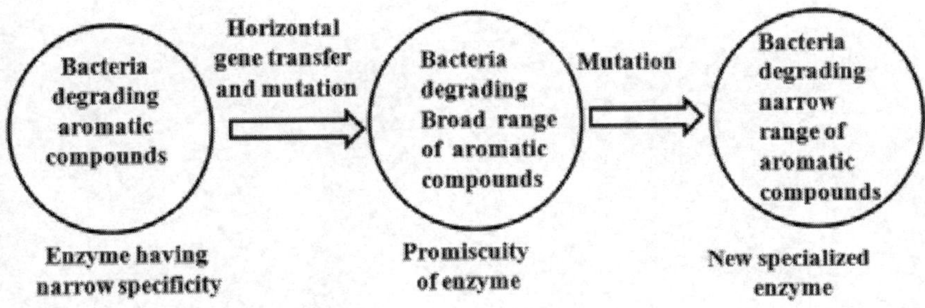

Figure 16. Evolution of new specialized enzymes from existing one. (Modified from Ju et al. [134])

The correct functioning of a pathway depends not only on having enzymes with appropriate catalytic activity but also on regulators, which control the expression of the catabolic genes in response to the compounds to be degraded. For example, ntdR, which controls 2NTDO expression in *Acidovorax* JS42, is supposed to have evolved from an ancestral nagR regulator of naphthalene degradation pathway present in *Ralstonia* U2. NtdR differs only by five amino acids with nagR. Ju et al. [134] showed how ntdR like regulator could be created from nagR by making mutations at each of the five positions separately and in combinations in a stepwise manner. They also showed that each mutation broadened the effectors range in a stepwise manner without losing the original activity. Both NagR and NtdR can activate transcription in the presence of salicylate, which is a natural inducer of naphthalene degradation genes in strain *Ralstonia* sp. U2, but ntdR could have gained a broader effector specificity to recognize several nitroaromatic compounds too [134]. Hence, the evolution of the regulatory system of the 2NTDO is in an intermediate stage because it can be induced in response to several nonmetabolizable compounds. Thus, the selection of ntdR variant with high specificity for 2NT with loss of specificity for salicylate would enable the identification of mutations that can lead to the specialized transcription factor from an intermediate stage. Another regulator was reported by Singh et al. [83], where the regulatory protein sequence of 3NTDO (MntR) differs only at three amino acid positions from NagR of *Ralstonia* sp. strain U2, out of which two are uniquely present in strain DS2 only.

It can be concluded that the gene evolution in these dioxygenase systems cannot be explained by considering only one mode of evolution. All the modes of evolution (like horizontal gene transfer, selective mutation, and promiscuity) are responsible for the evolution of a dioxygenase system [133, 135]. Further, the presence of truncated ORFs (which is not required for enzyme activity) reveals that gene evolution is in an intermediate stage of the so-called progressive compaction of the genes.

References

[1] Timmis K N, Steffan R J, Unterman R. Designing microorganisms for the treatment of toxic wastes. Annu Rev Microbiol. 1994;48:525–557.

[2] Faber K, Biotransformations in Organic Chemistry.4th ed. Springer; 2000.

[3] Alvarez P J J, Illman W A, Bioremediation and Natural Attenuation: Process Fundamentals and Mathematical Models. Hoboken, New Jersey: John Wiley & Sons, Inc.; 2006.

[4] Neilson A H, Allard A S. Environmental Degradation and Transformation of Organic Chemicals. Taylor & Francis Group; 2008.

[5] Ju K S, Parales R E, Nitroaromatic compounds, from synthesis to biodegradation. Microbiol Mol Biol Rev. 2010;74:250–272.

[6] Winkler R, Hertweck A. Biosynthesis of nitro compounds. Chem Bio Chem. 2007; 8:973–977.

[7] Ehrlich J, et al. *Streptomyces venezuelae*, n. sp., the source of chloromycetin. J Bacteriol. 1948; 56: 467–477.

[8] Gottlieb D, et al. Some properties of an antibiotic obtained from a species of *Streptomyces*. J Bacteriol. 1948;55: 409–417.

[9] Smith R M, et al. Chloromycetin: biological studies.J.Bacteriol. 1948, 55, 425–448.

[10] Donze G, McMahon C, Guerin P M. Rumen metabolites serve ticks to exploit large mammals. J Exp Biol, 2004;207:4283–4289.

[11] Atkinson R., et al. Kinetics and products of the gas-phase reactions of OH radicals and N2O5 with naphthalene and biphenyl. Environ Sci Technol, 1987;21:1014–1022.

[12] Atkinson, R. et al. Formation of ring-retaining products from the hydroxyl radical-initiated reactions of benzene and toluene.Int J Chem Kinet. 1989;21:801–827.

[13] Nishino N, Atkinson R, Arey J. Formation of nitro products from the gas-phase OH radical-initiated reactions of toluene, naphthalene, and biphenyl: effect of NO2 concentration. Environ Sci Technol, 2008;42:9203–9209.

[14] Purohit V, Basu A K, Mutagenicity of nitroaromatic compounds. Chem Res Toxicol, 2000;13: 673–692.

[15] Saitou, N. and K. Yamaguchi. Nitrobenzene and related compounds.In T. Shibamoto (ed.), Chromatographic Analysis of Environmental and Food Toxicants. CRC Press, Boca Raton, FL; 1998,169–212.

[16] Dalpozzo R, Bartoli G, Bartoli indole synthesis. Curr Org Chem. 2005;9:163–178.

[17] Bhattacharya, A., et al. One-step reductive amidation of nitro arenes: application in the synthesis of Acetaminophen.Tetrahedron Lett. 2006; 47:1861–1864.

[18] Schmidt, M., et al. Synthesis and biochemical characterization of new phenothiazines and related drugs as MDR reversal agents.Arch Pharm (Weinheim). 2008;341: 624–638.

[19] Gritsenko A N,Ermakova Z I, Zhuravlev S V, Synthesis in the phenothiazine series. Chem Heterocycl Compd. 1968;6:1245–1246.

[20] Hirai K. Structural Evolution and Synthesis of Diphenyl Ethers, Cyclic Imides and Related Compounds. Springer; 1999.

[21] Fletcher J H. et al. Synthesis of parathion and some closely related compounds. J Am Chem Soc. 1950;72:2461–2464.

[22] Singerman G M. 2,3-Dihydro-2,2-dimethyl-7-benzo[b]thienyl n-methylcarbamate and use as an insecticide. U.S. patent 4,032,649;1977.

[23] Ware G W. The Pesticide Book. 4th ed. Fresno, CA: Thompson Publications; 1994.

[24] Travis AS. Manufacture and uses of the anilines: a vast array of processes and products. In Z. Rappoport (ed.), The Chemistry of Anilines, Part 1. New York, NY: John Wiley and Sons; 2007,715–782.

[25] Lee P. Explosives development and fundamentals of explosives technology. In: J.A. Zukas and W. Walters (eds.), Explosive Effects and Applications. Berlin, Germany: Springer; 2002,23–43.

[26] Oxley J C. The Chemistry of Explosives. Springer; 1998,137–172.

[27] Spain J C. Biodegradation of nitroaromatic compounds. Ann Rev Microbiol. 1995;49:523–555.

[28] Lague, D. China Blames Oil Company for Benzene Spill in River. New York Times; 2005.

[29] Sax I R, Lewis R J. Nitro-compounds of aromatic hydrocarbons. Dangerous Properties of Industrial Material, 7th ed., vol. II; 1999,2534–2536.

[30] Haghighi-Podeh M R, Bhattacharya S K. Fate and toxic effects of nitrophenols on anaerobic treatment systems. Water Sci Technol. 1996;34:345–350.

[31] Talmage S S. et al. Nitroaromatic munition compounds: environmental effects and screening values.Rev Environ Contam Toxicol. 1999;161:1–156.

[32] Rickert D E, Long R M. Metabolism and excretion of 2,4-dinitrotoluene in male and female Fischer 344 rats after different doses. Drug Metab Dispos. 1981,9,226–232.

[33] Ilvicky J, Casida J E. Uncoupling action of 2,4-dinitrophenols, 2-trifluromethylbenzimidazoles and certain other pesticide chemicals upon mitochondria from different sources and its relation to toxicity.Biochem Pharmacol. 1969;18:444–445.

[34] Dodard S G. et al. Ecotoxicity characterization of dinitrotoluenes and some of their reduced metabolites. Chemosphere. 1999;38:2071–2079.

[35] Yen J H, Lin K H, Wang Y S. Acute lethal toxicity of environmental pollutants to aquatic organisms. Ecotoxicol Environ Saf. 2002;52:113–116.

[36] Burns LA., et al. Immunotoxicity of mono-nitrotoluenes in female B6C3F1 mice: I. Para-nitrotoluene. Drug Chem Toxicol. 1994;17:317–358.

[37] Rickert D E, Chism J P, Kedderis G L. Metabolism and carcinogenicity of nitrotoluenes.Adv Exp Med Biol. 1986;197:563–571.

[38] Spanggord R J. et al., Mutagenicity in *Salmonella typhimurium* and structure–activity relationships of wastewater components emanating from the manufacture of trinitrotoluene.Environ Mutagen. 1982;4:163–179.

[39] Won W D, DiSalvo L H, Ng J. Toxicity and mutagenicity of 2,4,6-trinitrotoluene and its microbial metabolites.Appl Environ Microbiol. 1976;31:576–580.

[40] Dunnick J K, Elwell M R, Bucher J R. Comparative toxicities of o-, m-, and p-nitrotoluene in 13-week feed studies in F344 rats and B6C3F1 mice. Fundam Appl Toxicol, 1994;22:411–421.

[41] Rickert D E, Butterworth B E, Popp J A. Dinitrotoluene: acute toxicity, oncogenicity, genotoxicity, and metabolism. Crit Rev Toxicol. 1984;13:217–234.

[42] Decad G M, Coraichen M E, Dent J D. Hepatic microsomal metabolism and covalent binding of 2,4-dinitrotoluene.Toxicol Appl Pharmacol. 1982;62:325.

[43] Rafii F. et al. Reduction of nitro-aromatic compounds by anaerobic bacteria isolated from the human gastrointestinal tract. Appl Environ Microbiol. 1991;57:962.

[44] Guittonneau S. et al. Oxidation of parachloronitrobenzene in dilute aqueous solution by [O3] + UV and [H2O2] + UV: a comparative study. Ozone Sci Eng. 1990;12:73–94.

[45] Kuo C H, Zappi M E, Chen S M. Peroxone oxidation of toluene and 2,4,6-trinitrotoluene.Ozone Sci Eng. 2000;22:519–534.

[46] Bin A K. et al. Degradation of nitroaromatics (MNT, DNT, and TNT) by AOPs.Ozone Sci Eng. 2001;23:343–349.

[47] Piccinini P. et al. Photocatalytic mineralization of nitrogen-containing benzene derivates. Catal Today. 1997;39:187–195.

[48] Kanekar P, Doudpure P, Sarnaik S. Biodegradation of nitroexplosives. Indian J Exp Biol. 2003;41:991–1001.

[49] Kulkarni M, Chaudhari A. Microbial remediation of nitro-aromatic compounds: an overview. J Environ Manage. 2007;85:496–512.

[50] Ye J, Singh A, Ward O P. Biodegradation of nitroaromatics and other nitrogen containing xenobiotics. World J Microbiol Biotechnol. 2004;20:117–135.

[51] Williams P A, Sayers J R. The evolution of pathways for aromatic hydrocarbon oxidation in *Pseudomonas*. Biodegradation. 1994;5:335–352.

[52] Dagley S. Biochemistry of aromatic hydrocarbon degradation in *Pseudomonas*. Gansalus I C, Sokatch J R, Ornston L N (eds.), The Bacteria. New York: Academic Press; 1986, 527–556.

[53] Harayama S, Timmis K N. Aerobic biodegradation of aromatic hydrocarbons by bacteria. Sigel H, Sigel A (eds.), Metal Ions in Biological Systems,Degradation of Environmental Pollutants by Microorganisms and Their Metalloenzymes. New York: Marcel Dekker Inc.; 1992, 99–156.

[54] Nishino S F, Spain J C, He Z. Biodegradation, transformation and bioremediation of nitroaromatic compounds. Hurst C J, Crawford R L, Knudsen G R, McInerey M J, Stetzenbach L D (eds.), Manual of Environmental Microbiology, 2nd ed. Washington, DC: ASM Press; 2002,987–996.

[55] Zhang C, Bennett G N. Biodegradation of xenobiotics by anaerobic bacteria. Appl Environ Microbiol. 2005;67:600–618.

[56] Razo-Flores E. et al. Biotransformation and biodegradation of N-substituted aromatics in methanogenic granular sludge. FEMS Microbiol Review. 1997;20:525–538.

[57] McCormick N, Feeherry F E, Levinson H S. Microbial transformation of 2,4,6-trinitrotoluene and other nitroaromatic compounds. Appl Environ Microbiol. 1976;31:949–958.

[58] Donlon B A, Razo-Flores E, Lettinga G, Field J A. Continuous detoxification, transformation and degradation of nitrophenols in upflow anaerobic sludge blanket (USAB) reactors. Biotechnol Bioeng. 1996;51:439–449.

[59] Peres C M. et al. Continuous degradation of mixtures of 4-nitrobenzoate and 4-aminobenzoate by immobilized cells of *Burkholderia cepacia* strain PB4. Appl Microbiol Biotechnol. 1999;52:440–445.

[60] Rieger P G, Knackmuss H J. Basic knowledge and perspectives on biodegradation of 2,4,6 trinitrotoluene and related nitroaromatic compounds in contaminated soil. In: J.C. Spain (ed.), Biodegradation of Nitroaromatic Compounds. New York: Plenum Press, 1995,1–18.

[61] Nishino S F, Spain J C, He Z. Strategies for aerobic degradation of nitroaromatic compounds by bacteria: process discovery to field application. Spain J C, Hughes J B, Knackmuss H J (eds.), Biodegradation of Nitroaromatic Compounds and Explosives. New York: Lewis Publ; 2000,7–61.

[62] Simpson J R, Evans W C. The metabolism of nitrophenols by certain bacteria. Biochem J. 1953;55:XXIV.

[63] Ecker S. et al. Catabolism of 2,6-dinitrophenol by *Alcaligenes eutrophus* JMT134 and 222. Arch Microbiol, 1992;158:149–154.

[64] Haigler B E, Wallace W H, Spain J C. Biodegradation of 2-nitrotoluene by *Pseudomonas* sp. strain JS42. Appl Environ Microbiol. 1994,60,3466–3469.

[65] Mulla S I. et al. Biodegradation of 2-nitrotoluene by *Micrococcus* sp.strain SMN-1. Biodegradation. 2011;22:95–102.

[66] Singh D, Ramanathan G. Biomineralization of 3-nitrotoluene by *Diaphorobacter* species. Biodegradation. 2013;24:645–655.

[67] Nishino S F, Spain J C. Oxidative pathway for the biodegradation of nitrobenzene by *Comamonas* sp. strain JS765. Appl Environ Microbiol. 1995;61:2308–2313.

[68] Suen W C, Spain J C. Cloning and characterization of *Pseudomonas sp*. strain DNT genes for 2,4-dinitrotoluene degradation. J Bacteriol. 1993;175:1831–1837.

[69] Groenewegen P E J, De Bont J A M. Degradation of 4-nitrobenzoate via 4-hydroxylaminobenzoate and 3,4-dihydroxybenzoate in *Comamonas acidovorans* NBA-10. Arch Microbiol. 1992;158:381–386.

[70] Ali-Sadat S, Mohan K S, Walia S K. A novel pathway for the biodegradation of 3-nitrotoluene in *Pseudomonas putida*. FEMS Microbiol Ecol. 1995;17:169–176.

[71] Williams R W, Taylor S C, Williams P A. A novel pathway for the catabolism of 4-nitrotoluene by *Pseudomonas*. J Gen Microbiol, 1993;139:1967–1972.

[72] Spiess T. et al. A new 4-nitrotoluene degradation pathway in a *Mycobacterium* strain. Appl Environ Microbiol. 1998;64:446.

[73] An D, Gibson D T, Spain J C. Oxidative release of nitrite from 2-nitrotoluene by a three-component enzyme system from *Pseudomonas sp.*strain JS42. J Bacteriol. 1994;176:7462–7467.

[74] Jain R K, Dreisbach J H, Spain J C. Biodegradation of p-nitrophenol via 1,2,4-benzenetriol by *Arthrobacter sp*. Appl Environ Microbial. 1994;60:3030–3032.

[75] Spain J C, Gibson D T. Pathway for biodegradation of p-nitrophenol in a *Moraxella sp.* Appl Environ Microbiol. 1991;57:812–819.

[76] Zeyer J, Kearney P C. Degradation of o-nitrophenol and m-nitrophenol by a *Pseudomonas putida*. J Agric Food Chem. 1984;32:238–242.

[77] Delgado A. et al. Nitroaromatics are substrates for the TOL plasmid upper-pathway enzymes. Appl Environ Microbiol. 1992;58:415–417.

[78] Robertson J B. et al. Oxidation of nitrotoluenes by toluene dioxygenase: evidence for a monooxygenase reaction. Appl Environ Microbiol. 1992;58:2643–2648.

[79] Struijs J, Stoltenkamp J. Ultimate biodegradation of 2-, 3-, and 4-nitrotoluene. Sci Total Environ. 1986;57:161–170.

[80] Goodall J L, Peretti S W. Dynamic modeling of meta- and para-nitrobenzoate metabolism by a mixed co-immobilized culture of *Comamonas sp*. JS46 and JS47. Biotechnol Bioeng, 1998;59:507–516.

[81] Williams R W, Taylor S C, Williams P A. A novel pathway for the catabolism of 4-nitrotoluene by *Pseudomonas*. J Gen Microbiol, 1993;139:1967–1972.

[82] Walia S K, Ali-Sadat S, Rasul Chaudhry G. Influence of nitro group on biotransformation of nitrotoluenes in *Pseudomonas putida* strain OU83.Pestic Biochem Physiol. 2003;76: 73–81.

[83] Singh D, Kumari A, Ramanathan G. 3-Nitrotoluene dioxygenase from *Diaphorobacter* sp. strains: cloning, sequencing and evolutionary studies. Biodegradation. 2014;25:479–492.

[84] Singh, D. et al. Expression, purification and substrate specificities of 3-nitrotoluene dioxygenase from *Diaphorobacter* sp. strain DS2. Biochem Biophys Res Commun. 2014;445:36–42.

[85] Spanggord R J. et al. Biodegradation of 2,4-dinitrotoluene by a *Pseudomonas sp*. Appl Environ Microbiol. 1991;57:3200–3205.

[86] Nishino S F, Paoli G C, Spain J C. Aerobic degradation of dinitrotoluenes and pathway for bacterial degradation of 2,6-dinitrotoluene. Appl Environ Microbiol. 2000;66:2139–2147.

[87] Nishino S F, Spain J C. Technology status review: bioremediation of dinitrotoluene (DNT), 2001

[88] Lendenmann U, Spain J C. Simultaneous biodegradation of 2,4-dinitrotoluene and 2,6-dinitrotoluene in an aerobic fluidizedbed biofilm reactor. Environ Sci Technol. 1998;32: 82–87.

[89] Bae B, Autenrieth R L, Bonner J S. Aerobic biotransformation and mineralization of 2,4,6-trinitrotoluene. In: R.E. Hinchee, R.E. Hoeppel, D.B. Anderson (eds.), Bioremediation of Recalcitrant Organics. Columbus, Ohio: Battelle Press; 1995,231–238.

[90] Paca J,Halecky M, Barta J, Bajpai R. Aerobic biodegradation of 2,4-DNT and 2,6-DNT: Performance characteristics and biofilm composition changes in continuous packed-bed bioreactors. J Hazard Mater. 2009;163:848–854.

[91] Wang Z Y, Ye Z F, Zhang M H. Bioremediation of 2,4-dinitrotoluene (2,4-DNT) in immobilized micro-organism biological filter. J Appl Microbiol. 2011;110:1476–1484.

[92] Lenke H, Achtnich C, Knackmuss H J. Perceptives of bioelimination of polynitroaromatic compounds. Spain J C, Highes J B, Knackmuss H J (eds.), Biodegradation of Nitroaromatic Compounds and Explosives. Boca Raton, FL: Lewis Publishers. 2000, 91–126.

[93] Huang S. et al. 2,4,6-Trinitrotoluene reduction by carbon monoxide dehydrogenase from *Clostridium thermoaceticum*. Appl Environ Microbiol. 2000;66:1474–1478.

[94] Van Aken B. et al. Transformation and mineralization of 2,4,6-trinitrotoluene (TNT) by manganese peroxidase from the white-rot basidiomycete *Phlebia radiata*. Biodegradation. 1999;10:83–91.

[95] French C E, Nicklin S, Bruce N C. Aerobic degradation of 2,4,6-trinitrotoluene by *Enterobacter cloacae* PB2 and by pentaerythritol tetranitrate reductase. Appl Environ Microbiol. 1998;64:2864–2868.

[96] Haidour A, Juan L R. Identification of products resulting from the biological reduction of 2,4,6-trinitrotoluene, 2,4-dinitrotoluene, and 2,6-dinitrotoluene by *Pseudomonas sp*. Environ Sci Technol. 1996;30:2365–2370.

[97] Shim C Y, Crawford D L. Biodegradation of trinitrotoluene (TNT) by a strain of *Clostridium bifermentans*. Hinchee R E, Fredrickson J, Alleman B C (eds.), Bioaugmentation for Site Remediation. Columbus, Ohio: Battelle Press; 1995,57–69.

[98] Lewis T A. et al. Microbial transformation of 2,4,6-trinitrotoluene. J Ind Microbiol Biotechnol. 1997;18:89–96.

[99] Boopathy R, Kulpa C F. Trinitrotoluene as a sole nitrogen source for a sulfate-reducing bacterium *Desulfovibrio sp.* (B strain) isolated from an anaerobic digester. Curr Microbiol. 1992;25:235–241.

[100] Preuss A, Fimpel J, Dickert G. Anaerobic transformation of 2,4,6-trinitrotoluene (TNT). Arch Microbiol. 1993;159:345–353.

[101] Drzyzga O. et al. Mass balance studies with 14C-labeled 2,4,6-trinitrotoluene (TNT) mediated by an anerobic *Desulfovibrio* species and an aerobic *Serratia* species. Curr Microbiol. 1998;37:380–386.

[102] Boopathy R, Kulpa C F. Biotransformation of 2,4,6-trinitrotoluene (TNT) by a *Methanococcus* sp. (strain B) isolated from a lake sediment. Can J Microbiol. 1994;40: 273–278.

[103] Esteve-Núñez A, Ramos J L. Metabolism of 2,4,6-trinitrotoluene by *Pseudomonas* sp. JLR11. Environ Sci Technol. 1998;32:3802–3808.

[104] Kalafut T. et al. Biotransformation patterns of 2,4,6-trinitrotoluene by aerobic bacteria. Curr Microbiol. 1998;36:45–54.

[105] French C E, Nicklin S, Bruce N C. Aerobic degradation of 2,4,6-trinitrotoluene by *Enterobacter cloacae* PB2 and by pentaerythritol tetranitrate reductase. Appl Environ Microbiol. 1998;64:2864–2868.

[106] Alvarez M A. et al. *Pseudomonas aeruginosa* strain MA01 aerobically metabolizes the aminodinitrotoluenes produced by 2,4,6-trinitrotoluene nitro group reduction. Can J Microbiol. 1995;41:984–991.

[107] Pak J W. et al. Transformation of 2,4,6-trinitrotoluene by purified xenobiotic reductase B from *Pseudomonas fluorescens* I-C. Appl Environ Microbiol. 2000;66:4742–4750.

[108] Naumova R P, Selivanovskaya S L U, Mingatina F A. Possibilities for the deep bacterial destruction of 2,4,6-trinitrotoluene. Mikrobiologia. 1988;57:218–222.

[109] Gilcrease C P, Murphy V G. Bioconversion of 2,4-diamino-6-nitrotoluene to a novel metabolite under anoxic and aerobic conditions. Appl Environ Microbiol. 1995;61:4209–4214.

[110] Fiorella P D, Spain J C. Transformation of 2,4,6-trinitrotoluene by *Pseudomonas pseudoalcaligenes* JS52. Appl Environ Microbiol. 1997;63:2007–2015.

[111] Vorbeck C. et al. Initial reductive reactions in aerobic microbial metabolism of 2,4,6-trinitrotoluene. Appl Environ Microbiol. 1998;64:246–252.

[112] Montpas S. et al. Degradation of 2,4,6-trinitrotoluene by *Serratia marcescens*. Biotechnol Lett. 1997;19:291–294.

[113] Collie S. et al. Degradation of 2,4,6-trinitrotoluene (TNT) in an aerobic reactor. Chemosphere. 1995;31:3025–3032.

[114] Cho Y S, Lee B U, Oh K H. Simultaneous degradation of nitroaromatic compounds TNT, RDX, atrazine, and simazine by *Pseudomonas putida* HK-6 in bench-scale bioreactors. J Chem Technol Biotechnol. 2008;83:1211–1217.

[115] Oh K, Kim Y. Degradation of explosive 2,4,6-trinitrotoluene by s-triazine degrading bacterium isolated from contaminated soil. Bull Environ Contam Toxicol, 1998;61:702–708.

[116] Gisi D, Stucki G, Hanselmann K W. Biodegradation of the pesticide 4,6-dinitro-ortho-cresol by microorganisms in batch cultures and in fixed-bed column reactors. Appl Microbiol Biotechnol. 1997;48:441–448.

[117] Labana S. et al. Pot and field studies on bioremediation of p-nitrophenol contaminated soil using *Arthrobacter protophormiae* RKJ100. Environ Sci Technol. 2005;39:3330–3337.

[118] Labana S. et al. A microcosm study on bioremediation of p-nitrophenol-contaminated soil using *Arthrobacter protophormiae* RKJ100. Appl Microbiol Biotechnol. 2005;68:417–424.

[119] Snellinx Z. et al. Biological remediation of explosives and related nitroaromatic compounds. Environ Sci Pollut Res Int. 2002;9:48–61.

[120] Qureshi A, Kapley A, Purohit H J. Degradation of 2,4,6-trinitrophenol (TNP) by *Arthrobacter* sp. HPC1223 isolated from effluent treatment plant. Indian J Microbiol. 2012;52:642–647.

[121] Van der Meer J R. et al. Molecular mechanisms of genetic adaptation to xenobiotic compounds. Microbiol Rev. 1992;56:677–694.

[122] van der Meer J R. Evolution of novel metabolic pathways for the degradation of chloroaromatic compounds.Antonie Van Leeuwenhoek, 1997;71:159–178.

[123] Ochman H, LawrenceJ G, Groisman E A. Lateral gene transfer and the nature of bacterial innovation. Nature. 2000;405:299–304.

[124] Koonin E V, Makarova K S, Aravind L. Horizontal gene transfer in prokaryotes: quantification and classification. Ann Rev Microbiol. 2001;55:709–742.

[125] Alexander M. Biodegradation and Bioremediation. 2nd ed. San Diego, CA: Academic Press; 1999.

[126] Tan H M. Bacterial catabolic plasmids. Appl Microbiol Biotechnol. 1999;51:1–12.

[127] Liu H. et al. Patchwork assembly of nag-like nitroarene dioxygenase genes and the 3-chlorocatechol degradation cluster for evolution of the 2-chloronitrobenzene catabo-

lism pathway in *Pseudomonas stutzeri* ZWLR2-1. Appl Environ Microbiol. 2011;77:4547–4552.

[128] Johnson G R, Jain R K, Spain J C. Origins of the 2,4-dinitrotoluene pathway. J Bacteriol. 2002;184:4219–4232.

[129] Fuenmayor S L. et al. A gene cluster encoding steps in the conversion of naphthalene to gentisate in *Pseudomonas sp*. strain U2. J Bacteriol. 1998;180:2522–2530.

[130] Shin K A, Spain J C. Pathway and evolutionary implications of diphenylamine biodegradation by *Burkholderia sp*. strain JS667. Appl Environ Microbiol. 2009;75:2694–2704.

[131] Copley S D. Enzymes with extra talents: moonlighting functions and catalytic promiscuity. Curr Opin Chem Biol. 2003;7:265–272.

[132] James L C, Tawfik D S. Conformational diversity and protein evolution- a 60-year-old hypothesis revisited. Trends Biochem Sci. 2003;28:361–368.

[133] Kivisaar M. Degaradation of nitroaromatic compounds: a model to study evolution of metabolic pathways. Mol Microbiol. 2009;74:777–781.

[134] Ju K S, Parales J V, Parales R E. Reconstructing the evolutionary history of nitrotoluene detection in the transcriptional regulator NtdR. Mol Microbiol. 2009;74:826–843.

[135] Kivisaar, M. Evolution of catabolic pathways and their regulatory systems in synthetic nitroaromatic compounds degrading bacteria. Mol Microbiol. 2011;82:265–268.

Chapter 3

Selection of Promising Bacterial Strains as Potential Tools for the Bioremediation of Olive Mill Wastewater

Abstract

The main objective of this paper was the selection of promising bacterial strains to be used as potential tools to remove phenols in olive mill wastewater (OMW) or in other food wastes. Therefore, 12 OMW samples were analyzed and 119 isolates were collected. After a preliminary screening on a medium containing vanillic and cinnamic acids, three isolates were selected to evaluate their viability in presence of different compounds (cinnamic, vanillic and caffeic acids, rutin, tyrosol and oleuropein) and a possible bioremediation effect. The isolates generally survived with phenols added and exerted a significant bioremediation activity in some samples (reduction of phenols by 20%). The last step was focused on the evaluation of the combined effects of pH, cinnamic and vanillic acids on the viability of a selected isolate (13M); the combination of the acids exerted a strong effect on the target, but alkaline pH played a protective role.

Keywords: Bioremediation, phenol degradation, phenolic compounds, olive mill wastewater

1. Introduction

Olive oil production is one of the most important food sectors in the Mediterranean area as olive processing is considered a traditional industry for its countries since ancient times [1]. It is mainly produced in Spain (36% of the global production), Italy (24%), Greece (17%) followed by Portugal, France, Cyprus, Croatia, Turkey, Syria, and Tunisia. New producers are Argentina, Australia, and South Africa.

Olives are processed through two methods: pressing (discontinuous process) and centrifuging (continuous process, two/three phase centrifugation). The main inconvenience of these methods is the production of a polluting by-product that is a dark effluent known as olive mill wastewater (OMW) [2].

The environmental impact of olive oil production is strong due to the use of large quantities of water and the production of OMW: 1,000 kg of olives produce 0.5 m^3–1.5 m^3 of OMWs [1]. They are by-products generally considered undesirable but inevitable for every olive processing.

As defined in reference [3], OMW is "a stable emulsion constituted by vegetation waters of the olives, water from the processing, olive pulp and oil." It is characterized by a particular color (intensive violet-dark brown up to black color), odor (strong specific olive oil smell), high degree of organic pollution (expressed as biological and chemical oxygen demand (BOD and COD) values), acidic pH, high electrical conductivity, high content of polyphenols, high buffer capacity, and high content of solid matter.

OMWs are generally composed of water (83%–96%) and organic fraction (3.5%–15%) composed of 1%–8% carbohydrates, 0.5%–2.4% nitrogen compounds, 0.5%–1.5% organic acids, 0.02%–1% of fatty acids such as propionic, butyric, etc., and 1%–1.5% of phenolic compounds consisting of a hydroxyl group (-OH) bound directly to an aromatic hydrocarbon group and pectins [4].

Concerning phenols, they comprise low molecular weight compounds and polyphenols. Low molecular weight compounds are represented by caffeic, cinnamic, 2,6-dihydroxybenzoic, *p*-hydroxybenzoic, syringic, 3,4,5-trimethoxybenzoic, vanillic, and veratric acids; they have phytotoxic effects and antibacterial activity. Polymeric phenols (lignins, tannins, etc.) cause the typical brownish-black color of OMW and are the most recalcitrant fraction of this effluent.

The quantitative and qualitative composition of OMWs are variable due to climatic conditions, variety, ripeness of olives, and extraction processes; generally, they are produced in high quantities in a short time, thus their disposal represents an important problem.

As OMWs are rich in nutrients they could be used to remediate arid or semi-arid regions but their phytotoxicy affect plant growth [5]. OMWs have the highest polluting rate within the food industry due to the fact that they are recalcitrant to traditional biodegradative methods. The reduction of COD and BOD values represent an important goal for many industries but the high content in phenols complicate waste management; they exert an antimicrobial activity towards wastewater microflora thus biodegradation is slowed [6].

For these reasons phenols are considered as undesirable compounds; thus, physical, chemical, or biological treatments are used to reduce their pollutant load.

Waste remediation has been traditionally performed through some expensive methods (incineration, pyrolysis, landfill, etc.). In recent years, the increasing trends towards green economy and friendly approaches for the environment are the background to design alternative ways. According to this point of view, numerous researchers proposed bioremediation, defined as "the process whereby organic wastes are biologically degraded under controlled conditions to an innocuous state or to level below the limits established by regulatory authorities" [7]. According to the Environmental Protection Agency (EPA), bioremediation is a "treatment that use naturally occurring organisms to break down hazardous substances into less toxic or nontoxic substances."

Thanks to their ubiquity and metabolic pathways (aerobic, anaerobic fermentation, and co-metabolism) microorganisms are able to degrade and utilize various toxic compounds as energy source. Generally, the aerobic biodegradation has a higher efficiency than anaerobic processes and it is widely used. Nevertheless, in many cases, aerobic and anaerobic processes can also be used in series to reduce the complexity or the toxicity of the contaminants.

Numerous bacteria such as *Bacillus pumilus* [8], *Pediococcus pentosaceus* [9], *Lactobacillus plantarum* [10], *Arthrobacter* sp. [11], *Azotobacter vinelandii* [12–14], *Azotobacter chroococcum* [15], *Pseudomonas putida*, and *Ralstonia* sp. [16, 17] were able to degrade and/or remove phenols from OMW.

Yeasts and molds are also able to degrade phenols, namely, *Candida tropicalis*, *Candida cylindracea* and *Yarrovia lypolitica* [18, 19, 3, 20, 4], and white-rot fungi such as *Phanerochaete chrysosporium* or the genus *Pleurotus* [21–25]. In addition, *Trametes versicolor*, *Funalia trogii*, *Lentinus edodes*, *Aspergillus niger*, and *Aspergillus terreus* have been also mentioned as phenol-degrading organisms [26]. The main objective of this paper was the selection of promising bacterial strains to be used as potential tools for bioremediation; namely, after the isolation of some strains from OMWs, they were studied in relation to their ability to grow in a medium containing two secondary phenols. Then, a validation on a lab scale was performed.

2. Materials and methods

2.1. Isolation and phenotyping of potential phenol-degrading strains

Twelve different samples of OMW were analyzed. Aliquots of 100 ml of each OMW sample were mixed with 900 ml of sterile Ringer solution (0.25×; Oxoid, Milan, Italy) and shaken at 100 rpm for 30 min at room temperature. Then, this homogenate was serially diluted with a sterile saline solution (0.9% NaCl) and plated onto a Plate Count Agar (PCA; Oxoid) at 30°C for 48–72 h (for mesophilic bacteria and *Bacillus*) and Pseudomonas Agar Base + Pseudomonas C-F-C supplement (Oxoid) at 25°C for 72 h for *Pseudomonadaceae*. The analyses were performed in duplicate over two different batches. From each batch, some colonies were randomly selected from plates, and stored at 4°C on Tryptone Soya Agar (TSA) slants (Oxoid, Milan, Italy). Phenotyping of isolates was carried out through different tests (Gram, catalase activity, oxidase test, proteolytic activity, and oxido-fermentation).

2.2. Selection of potential phenol-degrading strains

The ability of the isolates to grow with phenols added was evaluated using Mineral Salt Medium (MSM), a synthetic medium containing K_2HPO_4 (1.6 g/l), KH_2PO_4 (0.4 g/l), NH_4NO_3

(0.5 g/l), $MgSO_4*7H_2O$ (0.2 g/l), $CaCl_2$ (0.025 g/l), $FeCl_2$ (0.005 g/l), Agar (12 g/l), cinnamic ($C_9H_8O_2$) or vanillic acids ($C_8H_8O_4$) (0.5 or 1 g/l; Sigma-Aldrich, Milan, Italy). After streaking the isolates onto the surface of this modified medium, the plates were incubated at 30°C for mesophilic bacteria and *Bacillus* and at 25°C for *Pseudomonadaceae* for 72 h. MSM without phenolic compounds was used as control. For a second assay, the isolates were preliminary grown in MSM broth added of 0.025 g/l and 0.05 g/l of cinnamic and vanillic acids and incubated for 24 h; thereafter they were streaked onto the surface of MSM with phenols, as reported above.

2.3. Effect of phenolic compounds on the viable cell count

This step was performed on three selected strains (6P, 13M, and 44M); they were grown in TS broth incubated at 25°C (strain 6P) or at 30°C (strains 13M and 44M) for 48 h. Each culture was centrifuged at 4,000 rpm for 10 min and the pellet was re-suspended in sterile saline solution (0.9% NaCl); ca. 6–7 log cfu/ml were inoculated in MSM medium (1, 2, and 3 g/l), added with phenols (cinnamic acid; vanillic acid; caffeic acid-$C_9H_9O_4$; rutin hydrate-$C_{27}H_{30}O_{16}$ x H_2O; tyrosol-$C_8H_{10}O_2$; oleuropein-$C_{25}H_{32}O_{13}$; phenols were purchased from Sigma-Aldrich). MSM without phenols was used as control.

The samples were stored at 25°C–30°C for 33 days and periodically analyzed to evaluate the viable count on TSA and the content of phenols through the Folin-Ciocalteau method [27]. The analyses were performed in duplicate and the results analyzed through one-way Analysis of Variance (one-way ANOVA), using Tukey's test as the post-hoc comparison test, or t-student test for paired comparisons. The statistical analysis was performed using the software Statistica for Windows version 10.0 (Statsoft, Tulsa, OK, USA).

2.4. Combined effects of pH, cinnamic, and vanillic acids on the viable count of the strain 13M

The strain 13M was used as target; it was inoculated into MSM broth to 6–7 log cfu/ml. The amounts of cinnamic and vanillic acids and pH varied according to a 2^k experimental (Table 1). The samples were stored at 30°C and periodically analyzed to evaluate the viable count and phenol content (up to 12 days).

Combinations	pH	Cinnamic acid (g/l)	Vanillic acid (g/l)
A	7	0	0
B	7	0	2
C	7	2	0
D	7	2	2
E	9	0	0
F	9	0	2
G	9	2	0
H	9	2	2

Table 1. Combinations of 2^k design.

The analyses were performed in duplicate and the results of viable count analyzed through a multiple regression approach by using the option DoE/2^k design of the software Statistica for Windows.

3. Results

3.1. Isolation and screening on MSM

The viable count of mesophilic and spore-forming bacteria and *Pseudomonas* was fairly high (7–8 log cfu/ml) (Table 2); thus, we selected 119 isolates (46 labeled as mesophilic bacteria, 44 and 29 belonging to *Bacillus* and *Pseudomonas* genera, respectively).

	Samples											
	1	2	3	4	5	6	7	8	9	10	11	12
M	7.95	7.77	8.30	7.60	7.48	9.30	8.30	8.30	8.00	6.78	6.30	6.70
B	7.30	6.70	7.30	6.78	6.78	7.00	6.60	6.70	5.70	6.48	6.48	6.30
P	7.70	6.70	8.00	6.90	7.60	6.30	8.30	8.00	7.84	7.00	7.48	8.78

Table 2. Viable count (log cfu/g) of mesophilic bacteria (M), *Bacillus* (B), and *Pseudomonas* (P) in OMW samples. Data are the average (n=2).

The isolates were streaked on MSM with cinnamic or vanillic acids; Figure 1 shows the results for vanillic acid. Namely, 12 isolates were able to grow in presence of 0.5 g/l of this compound (Figure 1a) and 9 with 1g/l (Figure 1b). None of the isolates grew with cinnamic acid.

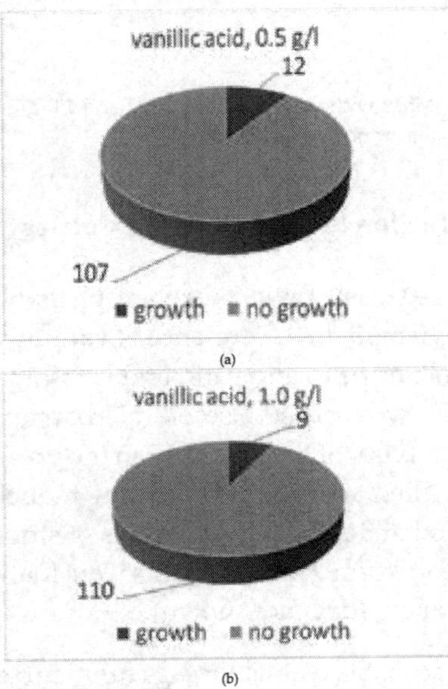

Figure 1. Screening of the isolates on MSM with vanillic acid (0.5 g/l and 1.0 g/l). The numbers on the pictures indicate if the targets are able or not able to grow.

The screening was also performed after isolate growing in MSM broth with low amounts of phenols; this step could be referred as an induction phase, aimed at inducing the resistance to phenols. Figure 2 shows the results with 1 g/l of vanillic acid. There were 32 out of 119 isolates that acquired the ability to grow in MSM with vanillic acid; at the lowest concentration (0.5 g/l) 49 strains were able to grow. The same protocol was also used for cinnamic acid, but only a single isolate was able to grow after induction both at 0.5 g/l and 1 g/l (the isolate 26M).

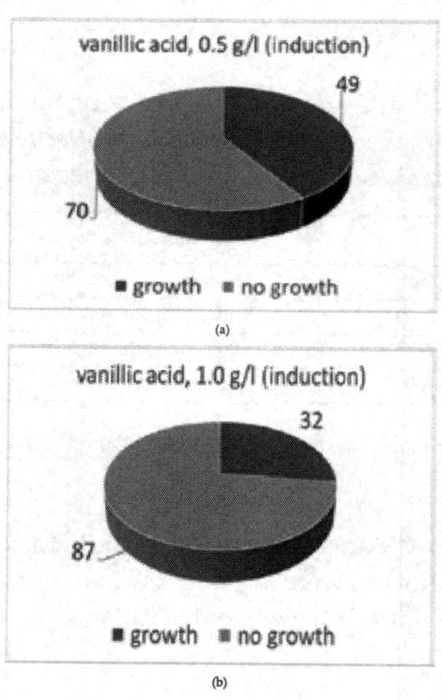

Figure 2. Screening of the isolates on MSM with vanillic acid (0.5 g/l and 1.0 g/l) after the induction. The numbers on the pictures indicate if the targets are able or not able to grow.

3.2. Viability and phenol reduction by some selected isolates

Three isolates (6P, 13M, and 44M; see Table 3 for their phenotypic traits) were selected and used as targets to assess their viability in the presence of various phenolic compounds (caffeic, cinnamic and vanillic acids, oleuropein, and rutin and tyrosol) at different concentrations (1 g/l, 2 g/l, and 3 g/l); moreover, we focused also on microorganism ability to reduce phenol content. Figure 3 reports the viability of the isolate 6P in the presence of caffeic acid; the initial cell number was 7 log cfu/ml. Then, it underwent a strong reduction within 5 days (ca. 1.5 log cfu/ml at 1 g/l and 2 log cfu/ ml at 3 g/l); in the last days of storage we found a tailing effect, with a residual cell count of 5 log cfu/ml. Similar results were found in the presence of cinnamic acid, tyrosol, rutin, and oleuropein (data not shown).

Vanillic acid at 2 g/l reduced the viable count by 3 log cfu/ml in 5 days with a final tailing effect and a residual cell count of 3–4 log cfu/ml. The lowest concentration (1 g/l) resulted in a slower death kinetic, with a similar residual viable count (Figure 4).

Phenotyping	Isolates		
	6P	13M	44M
Gram	-	+	+
Catalase	+	+	+
Oxydase	-	+	+
Proteolitic activity	-	+	+
O/F	F	F	F

Table 3. Phenotyping of the isolates selected for the second step of the research. F, metabolism under aerobic and anaerobic conditions.

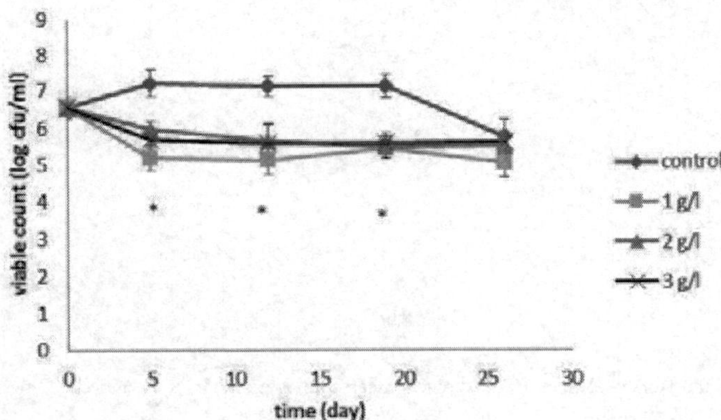

Figure 3. Viability of the isolate 6P in MSM+caffeic acid (mean values ± standard deviation). *, viable count in MSM +caffeic acid are significantly different from control.

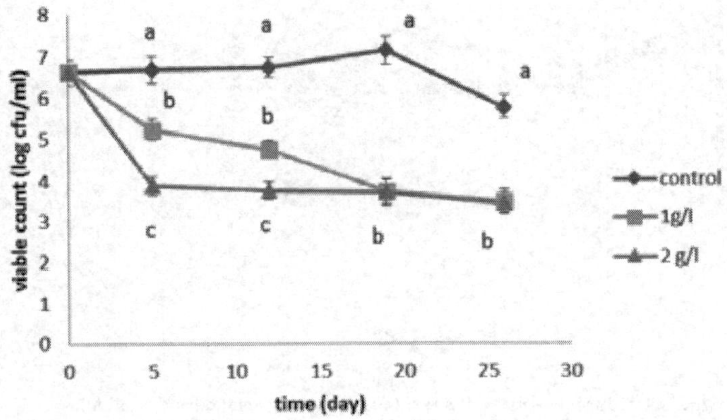

Figure 4. Viability of the isolate 6P in MSM+vanillic acid (mean values ± standard deviation). Letters indicate significant differences (one-way ANOVA and Tukey's test; P<0.05).

Phenols did not affect the viability of the strain 44M and the viable count was at 6–7 log/ml for the entire running time (data not shown). Some interesting results were found for the isolate 13M. Tyrosol at 3 g/l reduced the viable count by 3 log cfu/ml within 5 days, thereafter cell number increased up to 6–7 log cfu/ml (Figure 5); this trend could be the result of a kind of induction and adaptive evolution with phenols added. Caffeic acid at 1 g/l and 2 g/l caused a slight viability loss (1–2 log cfu/ml), while cell number was below the detection limit after 25 days at the highest concentration (3 g/l), thus suggesting a possible dose-dependent bactericidal effect (Figure 6). The other phenols did not affect the viable count (data not shown).

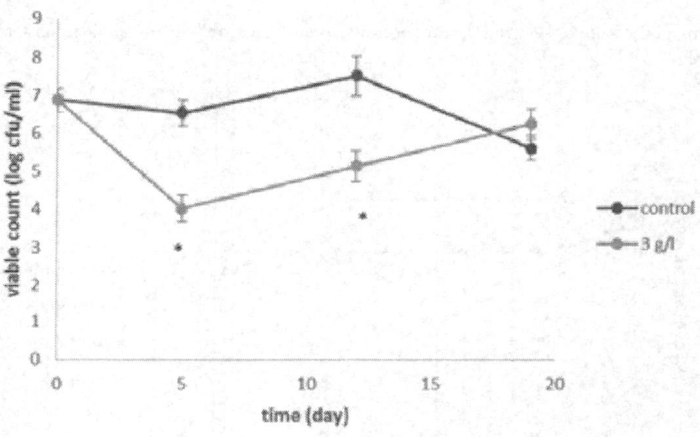

Figure 5. Viability of the isolate 13M in MSM+tyrosol (mean values ± standard deviation). *, significantly different from control (t-student test, $P<0.05$).

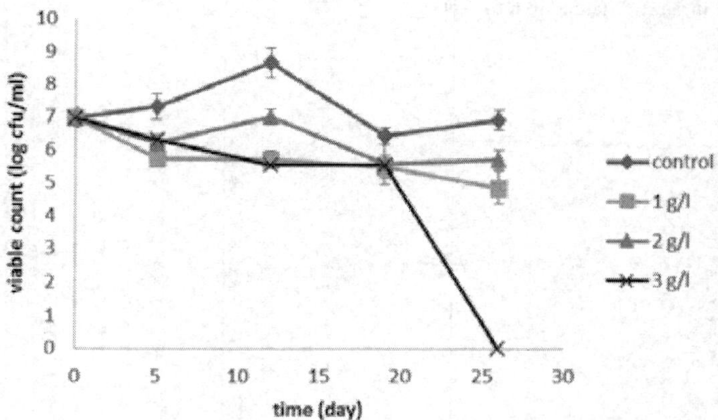

Figure 6. Viability of the isolate 13M in MSM+caffeic acid (mean values ± standard deviation).

We focused also on phenol content; Figure 7 shows the removal of vanillic acid by the isolate 13M and 44M (the initial content of the compound was 1 g/l). Both the strains were able to reduce its concentration in the broth, although the isolate 13M showed higher removal efficiency (ca. 18% after 33 days). The isolate 44M was also able to reduce cinnamic and caffeic acids by 24%–27% after 33 days (Figure 8). Removal was always found at 1g/l of phenols; higher initial amounts completely depleted bioremediation, probably due to a possible saturating action on the membrane.

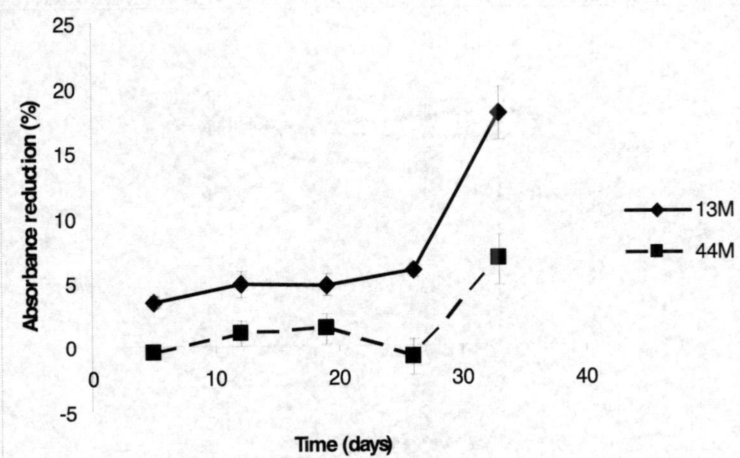

Figure 7. Removal of vanillic acid (initial content, 1 g/l) in MSM inoculated with the isolates 13M and 44 M; data are reported as absorbance fall. Mean values ± standard deviation.

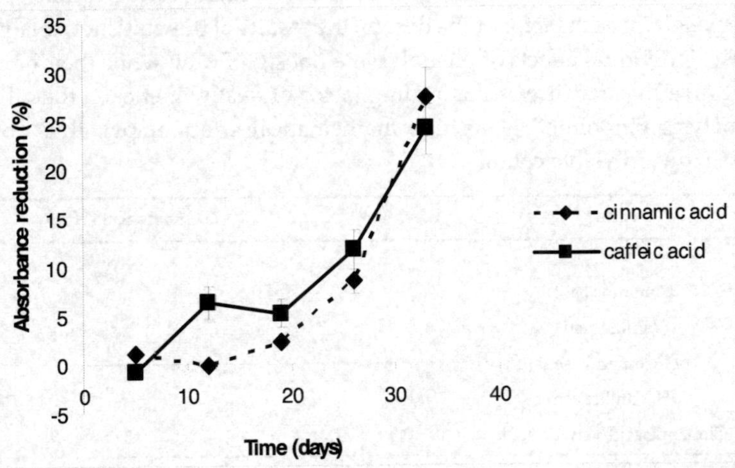

Figure 8. Removal of cinnamic and caffeic acids (initial content, 1 g/l) in MSM inoculated with the isolate 44 M; data are reported as absorbance fall. Mean values ± standard deviation.

3.3. Combined effects of cinnamic and vanillic acids and pH on cell count of the isolate 13M

This phase focused on the evaluation of pH, cinnamic and vanillic acids (combined through a 2^k experimental design) on the viability of the isolate 13M. Figure 9 shows the evolution of cell count in some selected combinations of the design; when MSM medium was adjusted to pH 7.0 in presence of 2 g/l of the phenols (combination D) the viable count (7 log cfu/ml) was drastically reduced to ca. 4.73 log cfu/ml. On the other hand, an alkaline pH played a protective role, as the viable count was not reduced both in absence and with phenol added (combinations E and H).

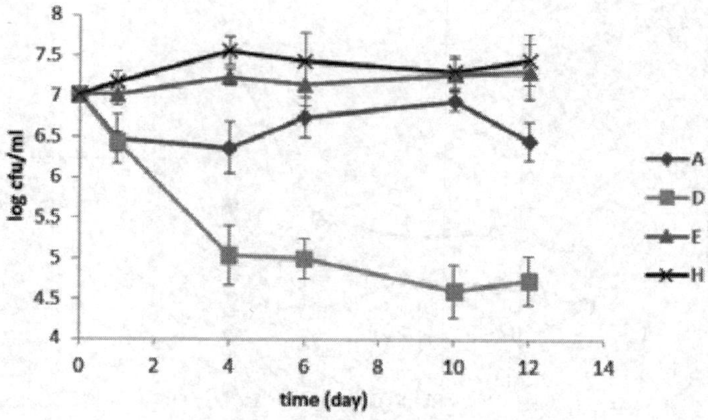

Figure 9. Viable count of the isolate 13M in some selected combinations of 2^k design (see Table 1).

Finally, Figure 10 shows the increase/decrease of cell count after 12 days; a positive value indicates a significant increase in the viable count, while a negative value indicates a death kinetic. These values were used as input data to run a multiple regression procedure and pinpoint the weight of each factor of the design; the results of this statistical analysis are listed in Table 4. The individual effects of phenols were not significant, while their interactive term played a negative role, i.e., it was the leading factor of death kinetic; on the other hand, the statistical analysis pinpointed a positive mathematical effect of pH, thus confirming its protective role toward viable count.

Terms	Statistical effect
pH	3.46
Cinnamic acid	ns
Vanillic acid	ns
pH/Cinnamic acid	ns
pH/Vanillic acid	ns
Cinnamic acid/vanillic acid	-3.06
R^2_{ad}	0.776

Table 4. Standardized effects of vanillic and cinnamic acids and pH on the reduction of the viable count of the isolate 13M after 12 days in MSM. Ns, not significant. R^2_{ad}, determination coefficient corrected for multiple regression.

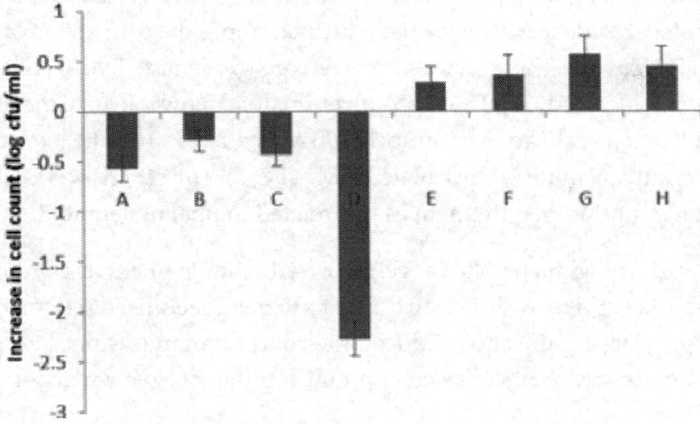

Figure 10. Increase/decrease of the viable count of the isolate 13M in MSM with phenols after 12 days. Mean values ± standard deviation (for the combinations see Table 1).

4. Discussion

The disposal of wastewaters represents a major problem. Namely, OMWs have the highest polluting rate within the food industry, especially for the high concentration of phenolic compounds. An increased interest in environmental issues has favored the introduction of new technologies as alternative ways to traditional methods. A promising approach is represented by bioremediation, which reduces the pollution load of various by-products using the phenol-degrading ability of some microorganisms. In a previous research, we selected some promising yeasts (*Pichia holstii* and *Candida boidinii*) able to reduce phenol content in OMW [26, 28]. Hereby, we evaluated bacterial ability to grow in a phenol-enriched medium; after isolation and phenotyping, we preliminary identified the most promising strains as *Bacillus* and *Pseudomonas* spp. These bacteria are normal constituents of wastewater microflora [29, 30]. After the screening with vanillic and cinnamic acids (chosen as representative of the most important phenolic moieties: coumaric and hydroxybenzoic compounds), we focused on viability in the presence of these compounds, as well as in a caffeic acid, rutin, oleuropein, and tyrosol enriched medium. Caffeic acid and rutin are representative of secondary phenolic compounds, while oleuropein is a high molecular weight phenolic glycoside responsible for the bitter taste in olive fruits. Oleuropein is present in olive mill wastewater as a result of debittering treatments and the extraction process of table and olive oil, respectively. Tyrosol is a product of hydrolysis of oleuropein. We used higher concentrations (1 g/l, 2 g/l, and 3 g/l) than those generally present in OMWs to evaluate bacterial viability in extreme conditions.

In some samples, phenols exerted a bactericidal action, due probably to their ability to form hydrogen bonds with proteins and/or enter cells. The resistant microorganisms generally convert phenols into carbolic acids; these latter compounds are transported through a transmembrane shuttle system into the cytoplasm. Carbolic acid is converted to catechol, and after three enzymatic steps, oxaloacetate is formed; the final step is the conversion of oxalacetate to acetaldehyde and pyruvate. The enzymes involved are oxygenase, hydroxylase, peroxidase, tyrosinase, and peroxidase [31, 32]. These products, finally, follow the main metabolic pathway up to their complete mineralization by mitochondrial chain [33, 34]. The oxalacetate can also be used in other cellular activities. The isolates 13M, 44M, and 6P strains were catalase positive and this trait is an important requirement as it is related to phenol degradation [28].

The isolates showed a good metabolic capacity towards simple phenolic compounds, as they were generally able to survive, with some exceptions to this generalized statement. In addition, some isolates (e.g., 13M) significantly reduced the concentration of some compounds in the broth and these are desired traits to select a promising microorganism acting as a bioremediation tool [35].

Concerning the second step of the research (combined effects of phenols and pH), the most important result was the effect of alkaline pH, as it seemed to exert a protective role on cell viability, thus we could suggest that phenolic metabolism at basic pHs is favored because the enzymes might have an optimal pH of 9. In these conditions a high presence of hydroxyl ions is ensured and it is very important as they represent a fundamental substrate used in the first step of the catabolic pathway of phenol to obtain catechol. These assumptions, however, require a confirmation. Finally, the protective effect of alkaline pH suggests the potential use of these isolates for the bioremediation of alkaline washing water of table olives.

5. Conclusions

Bioremediation could be considered as the promising solution for numerous food industry wastes, and to date, several works are in progress to isolate new phenol-degrading strains. This work concurs to confirm the importance of microorganisms to degrade pollutants; we selected some promising bacterial isolates, showing some desired traits in lab media. Further investigations are required to improve our work, i.e., evaluation of waste, evaluation of the effect on BOD and COD, and a focus on the role of alkaline pH on the removal.

References

[1] La Cara F, Ionata E, Del Monaco G, Marcolongo L, Gonçalves MR, Marques IP. Olive Mill Wastewater Anaerobically Digested: Phenolic Compounds with Antiradical Activity. Chemical Engineering Transactions. 2012; 27 325–330.

[2] Dias A, Bezerra RM, Periera AN. Activity and Elution Profile of Laccase During Biological Decolorization and Dephenolization of Olive Mill Wastewater. Bioresource Technology 2004; 92 (1) 7–13.

[3] Lanciotti R, Gianotti A, Baldi D, Angrisani R, Suzzi G, Mastrocola D, Guerzoni ME. Use of *Yarrowia lipolytica* Strains for the Treatment of Olive Mill Wastewaters. Bioresource Technology 2005; 96 (3) 317–322.

[4] Martinez-Garcia G, Johnson AC, Bachmann RT, Williams CJ, Burgoyne A, Edyvean RGJ. Anaerobic Treatment of Olive Mill Wastewater and Piggery Effluents Fermented with *Candida tropicalis*. Journal of Hazardous Materials 2009; 164 (1–2) 1398–1405.

[5] Paraskeva P, Diamadopoulos E. Technologies for Olive Mill Wastewater (OMW) Treatment: A Review. Journal of Chemical Technology and Biotechnology 2006; 81 (9) 1475–1485.

[6] Bevilacqua A, Augello S, De Stefano F, Campaniello D, Sinigaglia M, Corbo MR. Bioremediation of Wastes of Table Olives and Olive Oil: Environment-friendly Approaches. Industrie Alimentari 2014; 547 5–11.

[7] Mueller J.G., Cerniglia C.E., Pritchard P.H. Bioremediation of Environments Contaminated by Polycyclic Aromatic Hydrocarbons. In: Crawford R.L., Crawford D.L. (eds.) Bioremediation: Principles and Applications, Cambridge: Cambridge University Press 1996; 125–194.

[8] Ramos-Cormenzana A, Juarez-Jimenez B, Garcia-Pareja MP. Antimicrobial Activity of Olive Mill Wastewaters (Alpechin) and Biotransformed Olive Oil Mill Wastewater. International Biodeterioration and Biodegradation 1996; 38 (3–4) 283–290.

[9] Ben Othman N, Ayed L, Assas N, Kachouri F, Hammami M, Hamdi M. Ecological Removal of Recalcitrant Phenolic Compounds of Treated Olive Mill Wastewater by *Pediococcus pentosaceus*. Bioresource Technology 2008; 99 (8) 2996–3001.

[10] Ayed L, Hamdi M. Fermentative Decolorization of Olive Mill Wastewater by *Lactobacillus plantarum*. Process Biochemistry 2003; 39 (1) 59–65.

[11] Knupp G, Rucker G, Ramos-Cormenzana A, Hoyos SEG, Neugebauer M, Ossenkop T. Problems of Identifying Phenolic Compounds During the Microbial Degradation of Olive Mill Wastewater. International Biodeterioration and Biodegradation 1996; 38 (3–4) 277–282.

[12] Constantinos E, Papadopoulou K, Kotsou M, Mari I, Constantinos B. Adaptation and Population Dynamics of *Azotobacter vinelandii* During Aerobic Biological Treatment of Olive Mill Wastewater. FEMS Microbiology Ecology 1999; 30 (4) 301–311.

[13] Balis C, Chatzipavlidis J, Flouri F. Olive Mill Waste as a Substitute for Nitrogen Fixation. International Biodeterioration and Biodegradation 1996; 38 (3–4) 169–178.

[14] Piperidou CI, Chaidou CI, Stalikas D, Soulti K, Pilidis GA. Balis C. Bioremediation of Olive Mill Wastewater: Chemical Alterations Induced by *Azotobacter vinelandii.* Journal of Agricultural and Food Chemistry 2000; 48 (5) 1941–1948.

[15] Borja R, Martin A, Alonso V, Garcia I, Banks CJ. Influence of Different Aerobic Pretreatments on the Kinetics of Anaerobic Digestion of Olive Mill Wastewater. Water Research 1995; 29 (2) 489–495.

[16] Di Gioia D, Bertin L, Fava F, Marchetti L. Biodegradation of Hydroxylated and Methoxylated Benzoic, Phenylacetic and Phenylpropenoic Acids Present in Olive Mill Wastewaters by Two Bacterial Strains. Research Microbiology 2001; 152 (1) 83–93.

[17] Di Gioia D, Fava F, Bertin L, Marchetti L. Biodegradation of Synthetic and Naturally Occurring Mixtures of Mono-Cyclic Aromatic Compounds Present in Olive Mill Wastewaters by Two Aerobic Bacteria. Applied Microbiology and Biotechnology 2001; 55 (5) 619– 626.

[18] D'Annibale A, Brozzoli V, Crognale S, Gallo AN, Federici F, Petruccioli M. Optimisation by Response Surface Methodology of Fungal Lipase Production on Olive Mill Wastewater. Journal of Chemical Technology and Biotechnology 2006; 81 (9) 1586–1593.

[19] D'Annibale A, Sermanni GG, Federici F, Petruccioli M. Olive-Mill Wastewaters: A Promising Substrate for Microbial Lipase Production. Bioresource Technology 2006; 97 (15) 1828–1833.

[20] Martinez-Garcia G, Johnson AC, Bachman RT, Williams CJ, Burgoyne A, Edyvean RGJ. Two-Stage Biological Treatment of Olive Mill Wastewater with Whey as Co-substrate. International Biodeterioration and Biodegradation 2007; 59 (4) 273–282.

[21] Aloui F, Abid N, Roussos S, Sayadi S. Decolorization of Semisolid Olive Residues of "Alperujo" During the Solid State Fermentation by *Phanerochaete chrysosporium, Trametes versicolor, Pycnoporus cinnabarinus* and *Aspergillus niger*. Biochemical Engineering Journal 2007; 35 (2) 120–125.

[22] Garcìa Garcìa I, Jiménez Pena PR, Venceslada B, Martìn Martìn A, Martin Santos MA, Ramos Gomez E. Removal of Phenol Compounds from Olive Mill Wastewater Using *Phanerochaete chrysosporium, Aspergillus niger, Aspergillus terreus* and *Geotrichum candidum*. Process Biochemistry 2000; 35 (8) 751–758.

[23] Jaouani A, Sayadi S, Vanthournhout M, Penninckx MJ. Potent Fungi for Decolourisation of Olive Mill Wastewaters. Enzyme and Microbial Technology 2003; 33 (7) 802–809.

[24] Kissi M, Mountadar M, Assobhei O, Gargiulo E, Palmieri G, Giardina P, Sannita G. Roles of Two White-rot Basidiomyces Fungi in Decolorisation and Detoxification of Olive Mill Wastewater. Applied Microbial Biotechnology 2001; 57 (1-2) 221–226.

[25] Tsioulpas A, Dimou D, Iconomou D, Aggelis G. Phenolic Removal in Olive Oil Mill Wastewater by Strains of *Pleurotus* spp. in Respect to their Phenol Oxidase (Laccase) Activity. Bioresource Technology 2002; 84 (3) 251–257.

[26] Bevilacqua A., Petruzzi L., Corbo M.R., Sinigaglia M. Bioremediation of Olive Mill Wastewater by Yeasts – A Review of the Criteria for the Selection of Promising Strains. In: Patil Y. B., Rao P. (eds.) Applied Bioremediation - Active and Passive Approaches. Rijeka: Intech; 2013; 53–68.

[27] Bray H.G., Thorpe W.V. Analysis of phenolic compounds of interest in metabolism. In: Blick D. (ed.) Methods of Biochemical Analysis. New York: Interscience, 1954; 27–52.

[28] Sinigaglia M, Di Benedetto N, Bevilacqua A, Corbo MR, Capece A, Romano P. Yeasts Isolated from Olive Mill Wastewaters from Southern Italy: Technological Characterization and Potential Use for Phenol Removal. Applied Microbial Biotechnology 2010; 87 (6) 2345–2354.

[29] Ntougias S, Bourtzis K, Tsiamis G. The Microbiology of Olive Mill Wastes. BioMed Research International 2013; http://dx.doi.org/10.1155/2013/784591 (accessed 13 February 2015).

[30] Prakash B, Irfan M. *Pseudomonas aeruginosa* is Present in Crude Oil Contaminated Sites of Barmer Region (India). Journal of Bioremediation & Biodegradation 2011; 2:5 http://dx.doi.org/10.4172/2155-6199.1000129 (accessed 1 July 2014).

[31] Nordlund I, Powlowski J, Shingler V. Complete Nucleotide Sequence and Polypeptide Analysis of Multicomponent Phenol Hydroxylase from *Pseudomonas* sp. Strain CF600. Journal of Bacteriology 1990; 172 (12) 6826–6833.

[32] Nair CI, Jayachandran K, Shashidhar S. Biodegradation of Phenol. African Journal of Biotechnology 2008; 7 (25) 4951–4958.

[33] Zouari H, Moukha S, Labat M, Sayadi S. Cloning and Sequencing of a Phenol Hydroxylase Gene of *Pseudomonas pseudoalcaligenes* Strain MH1. Applied Biochemistry and Biotechnology 2002; 102–103 (1–6) 261–276.

[34] Kirchner U, Westphal AH, Müller R, van Berkel WJ. Phenol Hydroxylase from *Bacillus thermoglucosidasius* A7, a Two-protein Component Monooxygenase with a Dual Role for FAD. Journal of Biological Chemistry 2003; 278 (48) 47545–47553.

[35] Murínová S, Dercová K. Response Mechanisms of Bacterial Degraders to Environmental Contaminants on the Level of Cell Walls and Cytoplasmic Membrane. International Journal of Microbiology 2014; doi.org/10.1155/2014/873081.

Chapter 4

Gelation of Arabinoxylans from Maize Wastewater — Effect of Alkaline Hydrolysis Conditions on the Gel Rheology and Microstructure

Abstract

The purpose of this research was to extract arabinoxylans (AX) from maize wastewater generated under different maize nixtamalization conditions and to investigate the polysaccharide gelling capability, as well as the rheological and microstructural characteristics of the gels formed. The nixtamalization conditions were 1.5 hours of cooking and 24 hours of alkaline hydrolysis (AX1) or 30 minutes cooking and 4 hours of alkaline hydrolysis (AX2). AX1 and AX2 presented yield values of 0.9% and 0.5% (w/v), respectively. Both AX samples presented similar molecular identity (Fourier Transform Infra-Red) and molecular weight distribution but different ferulic acid (FA) content. AX1 and AX2 presented gelling capability under laccase exposure. The kinetics of gelation of both AX samples was rheologically monitored by small amplitude oscillatory shear. The gelation profiles followed a characteristic kinetics with an initial increase in the storage modulus (G') and loss modulus (G") followed by a plateau region for both gels. AX1 presented higher G' than AX2. In scanning electron microscopy (SEM) images, both gels present an irregular honeycomb microstructure. The lower FA content in AX2 form gels presenting minor elasticity values and a more fragmented microstructure. These results indicate that nixtamalization process conditions can modify the characteristics of AX gels.

Keywords: Ferulated arabinoxylans, nejayote, gelling, maize wastewater

1. Introduction

Nixtamalization is a process widely used in Mexico, the southern United States, Central America, Asia, and parts of Europe. This process consists of cooking maize grains in a lime solution, after soaking for 2–15 hours, the supernatant called maize wastewater or commonly known as "nejayote" is discarded [1, 2]. The remaining material is then ground to obtain nixtamal (dough or masa), used to prepare a variety of food products such as tortillas and related products [3]. A typical maize nixtamalization facility, processing 50 kg of maize every day, uses over 75 L of water per day and generates nearly the equivalent amount of alkaline wastewater on a daily basis [4]. Nejayote is considered an environmental pollutant because it is an alkaline wastewater, with high chemical and biological oxygen demand. The estimated monthly volume of nejayote generated in Mexico is about 1.2 m^3 [1, 5, 6]. Thus, alternatives for maize wastewater utilization are needed.

Nejayote is rich in maize bran residues as during the nixtamalization process this tissue is removed from the maize kernel. Non-starch polysaccharides are major constituents of maize bran, 30% of which are ferulated arabinoxylans (AX) [2, 7, 8]. Therefore, nixtamalization degrades and solubilizes maize cell wall components, mainly AX, which can be recovered in maize wastewater [4]. Nixtamalization conditions such as cooking temperature, lime concentration, and soaking period could affect the structural and functional characteristics of AX but, to our knowledge, the effect of maize nixtamalization conditions on AX properties has not been investigated.

AX are the main non-starchy polysaccharides of cereal grains, constituted of a linear backbone of β-(1→4)-linked D-xylopyranosyl units to which α-L-arabinofuranosyl substituents are attached through O-2 and/or O-3. Some of the arabinose residues are ester linked on (O)-5 to FA (3-methoxy, 4 hydroxy cinnamic acid) [9, 10]. AX can form covalent gels by oxidation of FA resulting in the formation of dimers (di-FA) and trimers (tri-FA) of FA as covalent cross-linking structures [11]. AX gels are stabilized by covalent linkages, which make them stable to temperature, pH or ionic strength changes; these characteristics would allow their passage through the gastrointestinal tract being further fermented by colonic microflora [12, 13]. In addition, AX gels could have potential application as a microencapsulation system for colon-specific delivery due to their porous structure (mesh size from 48 to 400 nm), high water absorption capacity (up to 100 g of water for gram of polymer), and dietary fiber nature [9, 10, 14].

The molecular features of AX depend on the source and the process extraction [10]. This characteristics as chemical structure, molecular weight and FA content affect their gelling ability and therefore functional properties of gels [11]. The purpose of this research was to extract AX from maize wastewater generated under two different maize nixtamalization conditions and investigate the polysaccharide gelling capability as well as the rheological and microstructural characteristics of the gels formed.

2. Experimental

2.1. Materials

Nejayote was provided by two tortilla-making in Northern Mexico and used to extract AX. Laccase, which is copper-containing oxidase enzyme (benzenediol:oxygen oxidoreductase, E.C.1.10.3.2) extracted from *Trametes versicolor* and the other chemical products were purchased from Sigma Co. (St. Louis, MO, USA).

2.2. Methods

2.2.1. Arabinoxylans extraction

AX from nejayote were extracted as previously described [15]. Maize wastewater generated under two commercial nixtamalization conditions were used: 1.5 hours of cooking and 24 hours of alkaline hydrolysis (AX1) or 0.5 hours of cooking and 4 hours of alkaline hydrolysis (AX2) (Scheme 1).

2.2.2. Fourier Transform Infra-Red (FT-IR) spectroscopy

FT-IR spectra of dry AX1 and AX2 were recorded on a Nicolet FT-IR spectrometer (Nicolet Instrument Corp. Madison, WI, US). The samples were pressed into KBr pellets (2 mg/200 mg KBr). A blank KBr disk was used as background. The spectra were measured in absorbance mode from 400–4000 cm^{-1} resolution [16].

2.2.3. Molecular weight distribution

Molecular weight distribution of AX1 and AX2 was determined by Size Exclusion-High Performance Liquid Chromatography (SE-HPLC) at 38°C using a TSKgel (Polymer Laboratories, Shropshire, UK) G500 PMWX column (7.8 x 300 mm). 20 μL of AX1 and AX2 solutions (0.5% w/v in 0.1 M $LiNO_3$) filtered through 0.2 μm (Whatman) were injected, and Water 2414 Refractive Index Detector was used for detection. Isocratic elution was performed at 0.6 mL/min with 0.1 M $LiNO_3$ filtered through 0.2 μm. Molecular weights were estimated after universal calibration with pullulans (polysaccharide extracted from the fermentation medium of the *Aureobasidium pullulans* consisting of maltotriose units) as standards (P50 to P800) [17].

2.2.4. Ferulic acid content

FA content in AX1 and AX2 were quantified by Reverse Phase High-Performance Liquid Chromatography (RP-HPLC) after de-esterification step as previously described [4, 18]. An Alltima C18 column (250 × 4.6 mm) (Alltech Associates, Inc., Deerfield, IL, USA) and a photodiode array detector Waters 996 (Millipore Co., Milford, MA, USA) were used. Detection was followed by UV absorbance at 320 nm. The measurements were performed in triplicate.

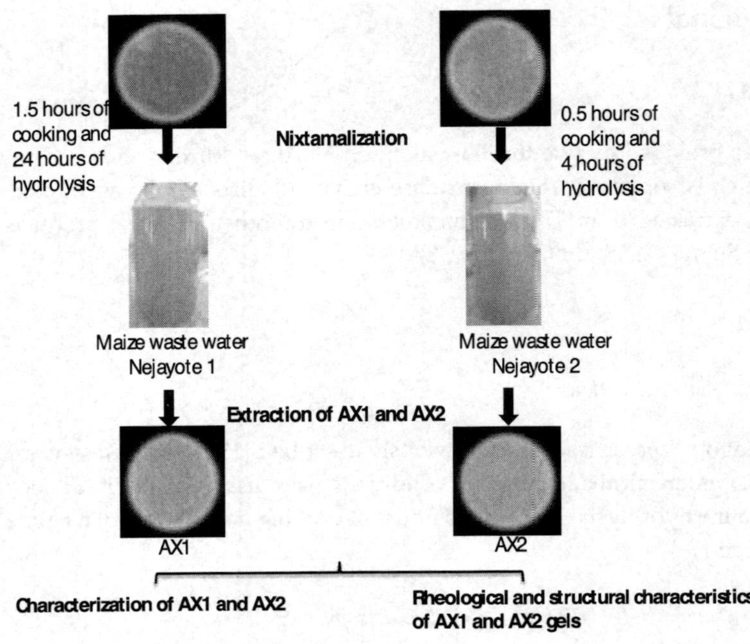

Scheme 1. Nixtamalization conditions and arabinoxylans extraction (AX1, AX2).

2.2.5. *Gel preparation*

AX1 and AX2 solutions at 10% (w/v) were prepared in 0.1 M sodium acetate buffer pH 5.5. Laccase (1.675 nkat per mg polysaccharide) was used as a cross-linking agent. AX1 and AX2 gels were allowed to develop for 4 hours at 25°C. The measurements were performed in duplicate.

2.2.6. *Rheological test*

Small amplitude oscillatory shear was used to follow the gelation process of AX1 and AX2 solutions at 10% (w/v). Solutions were mixed with laccase (1.675 nkat per mg AX) and immediately placed on the parallel plate geometry (4.0 cm in diameter) of a strain controlled rheometer (Discovery HR-3 rheometer; TA Instruments, New Castle, DE, US). Exposed edges were covered with silicone oil to prevent evaporation. The dynamic rheological parameters used to evaluate the gel network were the storage modulus (G'), loss modulus (G"), crossover point (G'>G"), and tan delta (tan δ, G"/G'). AX1 and AX2 gelation were monitored at 0.25 Hz and 5% strain. At the end of the network formation a frequency sweep (0.01–10 Hz) was carried out. Rheological measurements were performed in duplicate [4].

2.2.7. *Microstructure*

AX1 and AX2 gels at 10% (w/v) were frozen in liquid nitrogen and lyophilized at −37°C/0.133 mbar overnight in a Freezone 6 freeze drier (Labconco, Kansas, MO). The microstructure of the freeze-dried gels was studied by scanning electron microscopy (SEM) (JEOL 5410LV, JEOL, Peabody, MA, USA) at low voltage (20 kV) and Secondary Electron Imaging (SEI) mode [19].

2.2.8. Statistical analysis

FA content was made in triplicates and the coefficients of variation were lower than 5%. Small deformation measurements were made in duplicates and the coefficients of variation were lower than 5%. All results are expressed as mean values.

3. Results and discussion

3.1. Extraction and characterization of AX1 and AX2

AX1 and AX2 presented yield values of 0.9 and 0.5 % (w/v), respectively. This is consistent with a previous report where poorer AX yield of extraction were registered at lower times of alkaline hydrolysis [20]. Nevertheless, the AX yields found in the present study were smaller than those previously reported [21], which could be related to the maize varieties used in each investigation. The yield found in this study is similar to reported by other sources of AX as wheat flour (0.5 %) [16]. In spite of low yield values, recuperation of AX from wastewater could be an advantage for future industrial applications of this polysaccharide. It could also provide an alternative use this highly alkaline waste generated in large quantities.

The structural features of AX1 and AX2 were analyzed by FT-IR spectroscopy (Figure 1). The spectra were similar for both AX samples indicating a similar chemical structure characteristic of AX from maize and other sources [16, 17]. The region of 1200–850 cm^{-1} is typical of AX [17, 22, 23]. The maximum absorption band (~1035 cm^{-1}) could be assigned to C-OH bending with signals at 1,070 and 898 cm^{-1} that were related to the antisymmetric C-O-C stretching mode of the glycosidic bond and β(1-4) linkages between the xylose units [16, 17, 24]. Phenolic compounds and proteins have specific absorption bands in the 1500 – 1,700 cm^{-1} [23]. The region from 3500–1800 cm^{-1} is the fingerprint region of polysaccharides, with two bands (3,400 cm^{-1} corresponding to stretching of the OH groups and 2,900 cm^{-1} corresponding to the CH_2 groups) [16, 25]. These results suggest that nixtamalization conditions used in the present study do not affect the structural features of AX. However, in this work the effect of two alkaline treatments was also investigated on the FA content and physicochemical characteristics of AX1 and AX2.

The gel permeation chromatography profile of AX1 and AX2 are presented in Figure 2. Molecular weight distribution profiles were similar for both AX showing a major peak at molecular weight region of ~250 kDa (high molecular weight region), a similar behavior has been previously reported for maize AX [17].

FA content in AX1 and AX2 was $0.012 \pm 2.7 \times 10^{-5}$ and $0.008 \pm 1.4 \times 10^{-4}$ (μg/mg polysaccharide), respectively. These values are lower than those reported for other maize wastewater AX (0.23 μg/mg polysaccharide) [4]. Heating temperature, lime concentration, hydrolysis time, and exposure to light could have affected the amount of FA present in AX. In a previous report found that the best conditions of alkaline hydrolysis for FA extraction (from brewer's spent grain) are low NaOH concentration (2.0%), temperature of 120°C, and a short reaction time (90 minutes) [26]. In the present work, maize grains cooking in a lime solution was performed during 90 minutes and 30 minutes for AX1 and AX2, respectively; but after heating, long soaking periods were used (24 hours and 4 hours for AX1 and AX2, respectively), which could explain the lower FA content in the polysaccharide. This is congruent with a previous study where the FA content in AX was dependent of the time of alkaline hydrolysis [20].

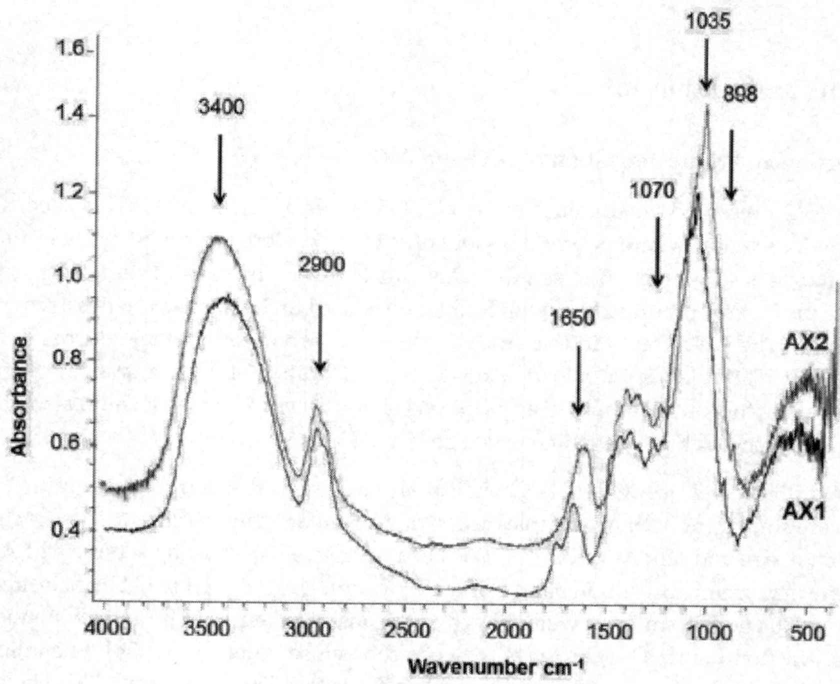

Figure 1. FT-IR of AX1 and AX2. The arrows indicate the characteristic absorption bands.

3.2. Rheological and structural characteristics of AX gels

AX1 and AX2 solutions at 10% (w/v) presented gelling capability under laccase exposure (Figure 3). The kinetics of gelation of these solutions was rheologically monitored by small amplitude oscillatory shear. Figure 4 shows the development of storage modulus (G'), loss (G") modulus, and tan δ (G"/G') versus time of 10% (w/v) AX1 and AX2 solutions undergoing oxidative gelation by laccase.

The gelation profiles followed a characteristic kinetics with an initial increase in G' and G", followed by a plateau region for both gels. This behavior reflects an initial formation of covalent linkages between FA of adjacent AX molecules producing a three-dimensional network [19]. At the end of gelation, G' and G" were 78 and 13 Pa for AX1, respectively, while for AX2 they were 32 Pa and 8 Pa for G' and G", respectively (Table 1). Similar kinetics of gelation have been

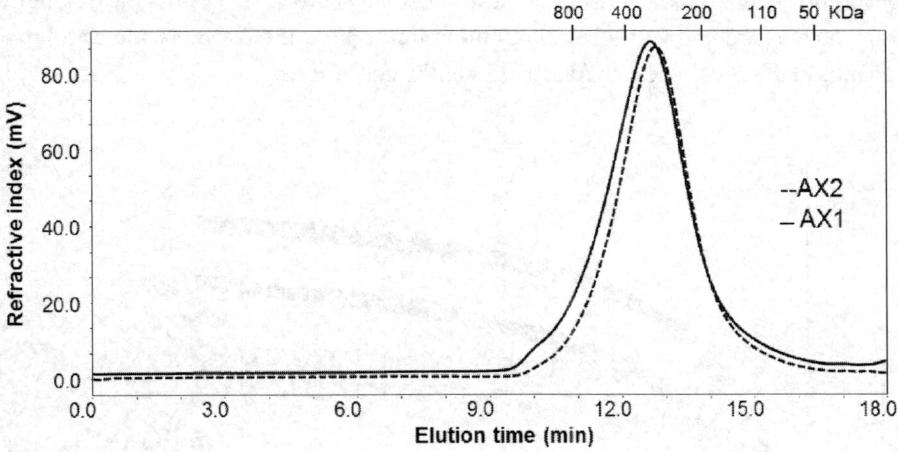

Figure 2. Elution profiles of AX1 and AX2. Pullulan molecular weight markers (kDa) used as calibration scales are shown at the top.

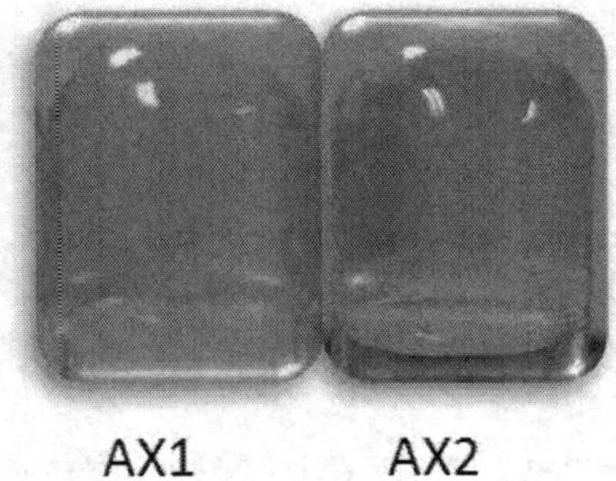

Figure 3. AX1 and AX2 gels at 10% (w/v).

previously reported for maize bran AX gels [17, 27, 28]. AX1 gel presented higher elasticity value in comparison to AX2 gel, which can be attributed to its higher FA content.

Gelation time (tg) at crossover point (G' > G") was 26 min and 40 min for AX1 and AX2, respectively. The tg value indicates the sol/gel transition point and at this point G' = G". The lower FA content in AX2 compared with AX1 could have affected the cross-linking of AX chains and retard the gel formation. The tan δ (G"/G') values decreased during the AX1 and AX2 gelation indicating the formation of a more elastic covalent system (Figure 4) [17]. The tan δ calculated at the end of the test, were 0.16 for AX1 and 0.24 for AX2, indicating than AX1 gel is more elastic than AX2 gel [29].

Niño-Medina et al. [21] reported nejayote AX gels (4% and 8%, w/v) with smaller G' values (2 and 4 Pa, respectively) and a higher crossover point (150 min) than those found in the present work. On the other hand, Ayala-Soto et al. [30] reported nejayote AX gels (4 % w/v) that showed a fluid-like behavior with G' of 5.9 Pa. Such differences might have its origin in the structural and/or conformational characteristics of these macromolecules [21, 31]. Possible differences in structure such as arabinose and FA distribution throughout the AX molecule could explain the variance in the rheological characteristics of the gels formed.

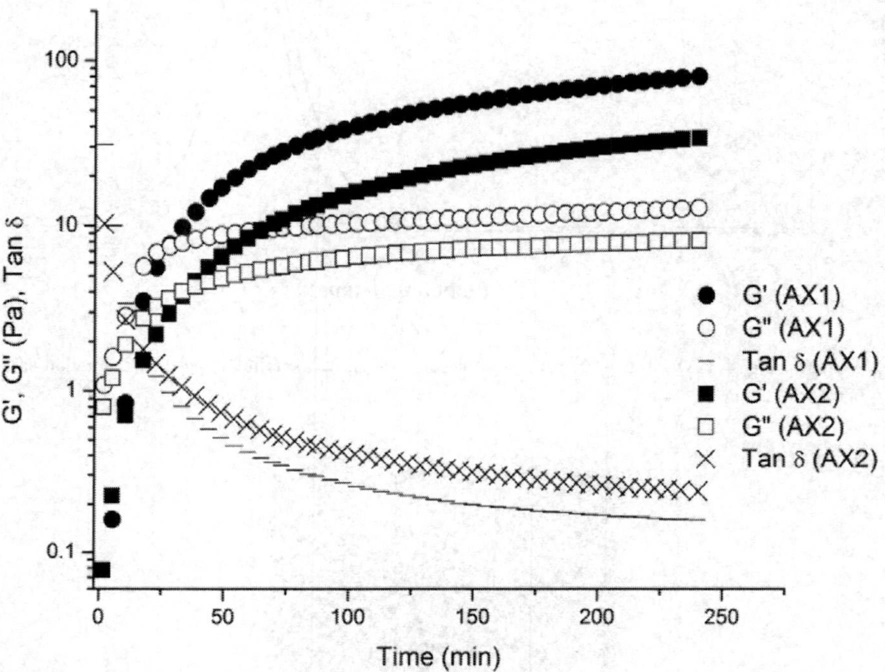

Figure 4. Monitoring the storage (G') and loss modulus (G") of AX1 and AX2 solutions (10% w/v) during gelation by laccase at 0.25 Hz and 25°C.

AX	Hydrolysis time (h)	Ferulic acid, FA (μg/mg AX)	Gelation time, tg (min)	G' (Pa)	G" (Pa)	Tan delta (δ, G"/G')
AX1	24	$0.012 \pm 2.7 \times 10^{-5}$	26	78	13	0.16
AX2	4	$0.008 \pm 1.4 \times 10^{-4}$	40	32	8	0.24

Table 1. FA content in AX1 and AX2 and rheological characteristics of the gels formed at 10% (w/v).

The mechanical spectra of AX1 and AX2 after gelation (Figure 5) exhibited a solid-like behavior with G'>G". The mechanical spectra of AX gels with a linear G' independent of frequency and G" much smaller than G' and dependent of frequency have been previously reported [11, 16, 17, 27]. AX1 and AX2 gels G' increase at frequency values, this may indicate the presence of physical interactions in the polymer network in addition to the covalent bonds induced by laccase.

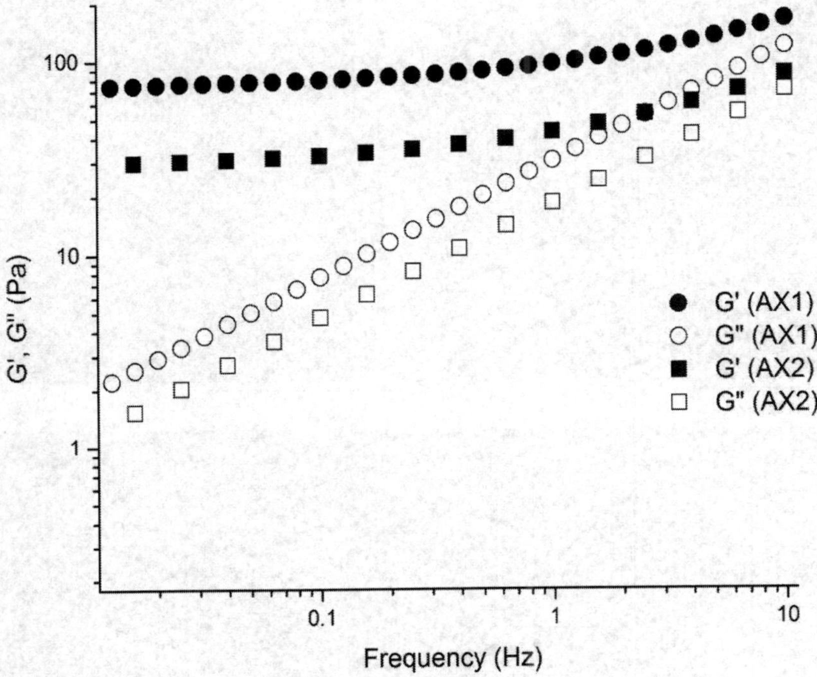

Figure 5. Mechanical spectra of AX1 and AX2 gels at 4 h. Rheological measurements made at 25 °C and 5% strain.

The images from SEM of the lyophilized gels AX1 and AX2 are shown in Figure 6. Both gels present many connections and resemble an imperfect honeycomb. In general, the microstructural characteristics of AX1 and AX2 are similar to those previously reported for lyophilized wheat and maize bran AX gels [16, 17, 22]. Nevertheless, AX2 gel appears to have a more fragmented morphology with a rougher and heterogeneous surface (Figure 6c, d). These microstructural dissimilarities between AX1 and AX2 gels could explain the differences in G' values of the gels, since a more compact and defined microstructure could give stronger gels.

The average inner diameter of the AX1 and AX2 cells were approximately 30 μm and 50 μm, respectively. Higher cell dimensions were reported in lyophilized AX gels (>200 μm) [17, 19], and this difference could be related to the method used to freeze the gels before lyophilization. In the present study AX gels were frozen by immersion in liquid nitrogen (fast congelation), while previous studies [17, 19] reported AX gel congelation at -20°C for several hours (slow congelation).

An important aspect of achieving a high-quality frozen material, particularly with high water content such as gels, is the freezing rate. Fast congelation results in a better-preserved structure (i.e., finer ice crystals). The microstructural characteristics of AX1 and AX2 gels were similar to those reported in AX gels frozen by fast congelation [32, 33] under similar conditions to those used in the present study. In those previous studies [32, 33], AX gels can be also compared with an irregular honeycomb structure in which pore diameters ranged from 12 μm to 50 μm.

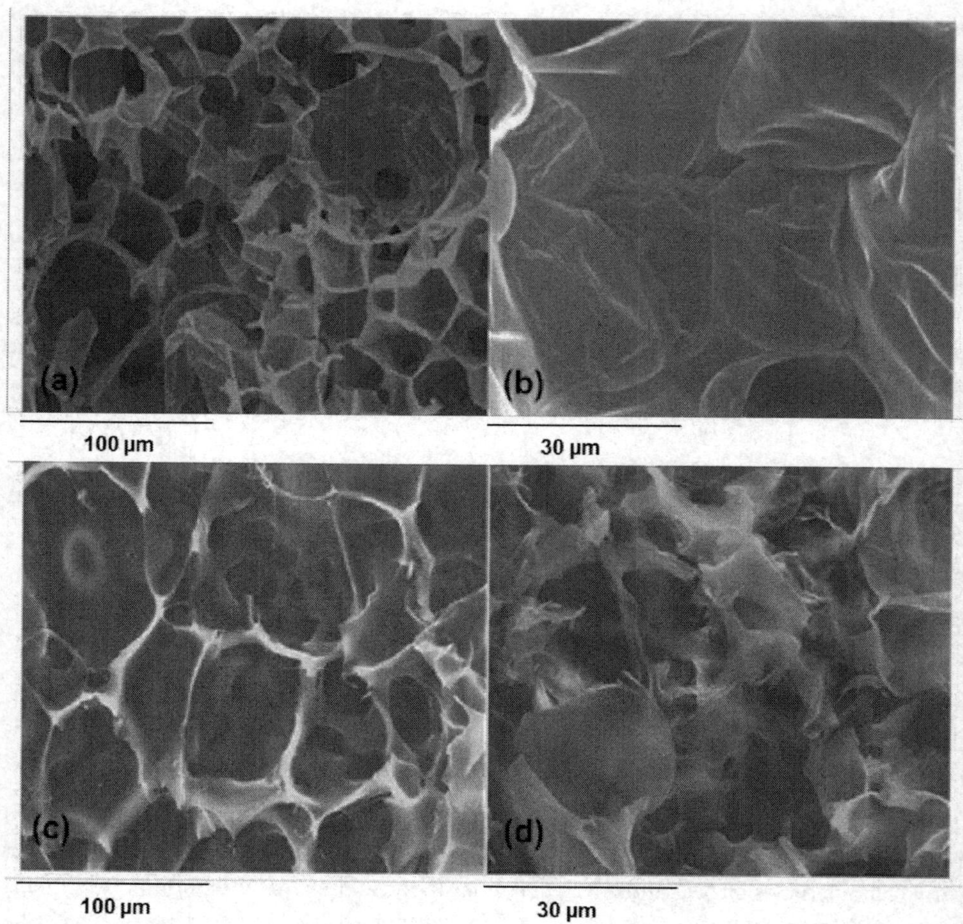

Figure 6. The SEM of lyophilized AX1 (a, b) and AX2 (c, d) gels. a and c at 500x magnification; b and d at 2000x magnification.

The entrapment of biomolecules or microorganisms in high cell dimension (>200μm) AX gels has been previously reported [14, 19, 31], but reducing cell dimensions of AX gels could increase the possibility of carried out smaller compounds or cells. Therefore, AX1 and AX2 gels microstructural characteristics could be of interest for the development of designed delivery systems, which could allow alternative uses for maize wastewater.

4. Conclusion

AX1 and AX2 presented similar molecular identity (FT-IR) and molecular weight distribution, but different FA content. Both AX1 and AX2 presented gelling capability under laccase exposure. The lower FA content in AX2 form gels presenting minor elasticity values and a more fragmented microstructure. These results indicate that nixtamalization process conditions can modify the characteristics of AX gels. Environmental concerns have triggered research on alternative nixtamalization processes rendering residues with less environment impact. AX recovering from this kind of less pollutant maize wastewater could be an interesting research subject in order to explore the structural and functional properties of this hydrocolloid.

References

[1] Valderrama-Bravo C, Gutiérrez-Cortez E, Contreras-Padilla M, Rojas-Molina I, Mosquera JC, Rojas-Molina A, Beristain F, Rodríguez-García, ME. Constant pressure filtration of lime water (nejayote) used to cook kernels in maize processing. Journal of Food Engineering. 2012;110(3):478-486. DOI: 10.1016/j.jfoodeng.2011.12.018.

[2] Acosta-Estrada BA, Lazo-Vélez MA, Nava-Valdez Y, Gutiérrez-Uribe JA, Serna-Saldívar SO. Improvement of dietary fiber, ferulic acid and calcium contents in pan bread enriched with nejayote food additive from white maize (Zea mays). Journal of Cereal Science. 2014;60(1):264-269. DOI: 10.1016/j.jcs.2014.04.006.

[3] Martínez-Bustos F, Martínez-Flores HE, Sanmartín-Martínez E, Sánchez-Sinencio F, Chang YK, Barrera-Arellano D, Rios E. Effect of the components of maize on the

quality of masa and tortillas during the traditional nixtamalisation process. Journal of the Science of Food and Agriculture. 2001;81:1455-1462. DOI: 10.1002/jsfa.963.

[4] Niño-Medina G, Carvajal-Millán E, Lizardi J, Rascon-Chu A, Marquez-Escalante JA, Gardea A, Martinez-Lopez AL, Guerrero V. Maize processing waste water arabinoxylans: Gelling capability and cross-linking content. Food Chemistry. 2009;115(4): 1286-1290. DOI: 10.1016/j.foodchem.2009.01.046.

[5] Gutiérrez-Uribe JA, Rojas-García C, García-Lara S, Serna-Saldivar SO. Phytochemical analysis of wastewater (nejayote) obtained after lime-cooking of different types of maize kernels processed into masa for tortillas. Journal of Cereal Science. 2010;52(3): 410-416. DOI: 10.1016/j.jcs.2010.07.003.

[6] Ayala-Soto FE, Serna-Saldívar SO, García-Lara S, Pérez-Carrillo E. Hydroxycinnamic acids, sugar composition and antioxidant capacity of arabinoxylans extracted from different maize fiber sources. Food Hydrocolloids. 2014;35:471-475. DOI: 10.1016/ j.foodhyd.2013.07.004.

[7] Martínez-López AL, Carvajal-Millan E, Lizardi-Mendoza J, López-Franco Y, Rascón-Chu A, Salas-Muñoz E, Ramírez-Wong B. Ferulated arabinoxylans as by-product from maize wet-milling process: characterization and gelling capability. In: Jimenez-Lopez JC, editor. Maize cultivation, uses and health benefits. New York: Nova Science Publisher, Inc.; 2012. p. 65–73.

[8] Saulnier L, Guillon F, Sado PE, Chateigner-Boutin AL, Rouau X. Plant cell wall polysaccharides in storage organs: Xylans (Food Applications). 2013. DOI: 10.1016/ b978-0-12-409547-2.01493-1.

[9] Izydorczyk M, Biliaderis C. Cereal arabinoxylans: advances in structure and physicochemical properties. Carbohydrate Polymers. 1995;28:33-48.

[10] Niño-Medina G, Carvajal-Millán E, Rascon-Chu A, Marquez-Escalante JA, Guerrero V, Salas-Muñoz E. Feruloylated arabinoxylans and arabinoxylan gels: structure, sources and applications. Phytochemistry Reviews. 2010;9(1):111-120. DOI: 10.1007/ s11101-009-9147-3.

[11] Carvajal-Millan E, Landillon V, Morel M-H, Rouau X, Doublier J-L, Micard V. Arabinoxylan gels: impact of the feruloylation degree on their structure and properties. Biomacromolecules. 2004;6(1):309-317. DOI: 10.1021/bm049629a.

[12] Hopkins MJ, Englyst HN, Macfarlane S, Furrie E, Macfarlane GT, McBain AJ. Degradation of cross-linked and non-cross-linked arabinoxylans by the intestinal microbiota in children. Applied and Environmental Microbiology. 2003;69(11):6354-6360. DOI: 10.1128/aem.69.11.6354-6360.2003.

[13] Grootaert C, Delcour JA, Courtin CM, Broekaert WF, Verstraete W, Van de Wiele T. Microbial metabolism and prebiotic potency of arabinoxylan oligosaccharides in the

human intestine. Trends in Food Science & Technology. 2007;18(2):64-71. DOI: 10.1016/j.tifs.2006.08.004.

[14] Carvajal-Millan E, Berlanga-Reyes C, Rascón-Chu A, Martínez-López AL, Márquez-Escalante JA, Campa-Mada AC, Martínez-Robinson KG. In vitro evaluation of arabinoxylan gels as an oral delivery system for insulin. Materials Research Society. 2012. DOI: 10.1557/opl.2012.

[15] Carvajal-Millán E, Rascón-Chu A, Márquez-Escalante JA. Método para la obtención de goma de maíz a partir del líquido residual de la nixtamalización del grano de maíz. Mexican Patent 2005;No. 278,768.

[16] Morales-Ortega A, Carvajal-Millan E, López-Franco Y, Rascón-Chu A, Lizardi-Mendoza J, Torres-Chavez P, Campa-Mada A. Characterization of water extractable arabinoxylans from a spring wheat flour: rheological properties and microstructure. Molecules. 2013;18(7):8417-8428. DOI: 10.3390/molecules18078417.

[17] Martínez-López AL, Carvajal-Millan E, Rascón-Chu A, Márquez-Escalante J, Martínez-Robinson K. Gels of ferulated arabinoxylans extracted from nixtamalized and non-nixtamalized maize bran: rheological and structural characteristics. CyTA - Journal of Food. 2013;11(sup1):22-28. DOI: 10.1080/19476337.2013.781679.

[18] Vansteenkiste E, Babot C, Rouau X, Micard V. Oxidative gelation of feruloylated arabinoxylan as affected by protein. Influence on protein enzymatic hydrolysis. Food Hydrocolloids. 2004;18(4):557-564. DOI: 10.1016/j.foodhyd.2003.09.004.

[19] Morales-Ortega A, Carvajal-Millan E, Brown-Bojorquez F, Rascón-Chu A, Torres-Chavez P, López-Franco Y, Lizardi-Mendoza J, Martínez-López AL, Campa-Mada AC. Entrapment of probiotics in water extractable arabinoxylan gels: rheological and microstructural characterization. Molecules. 2014;19(3):3628-3637. DOI: 10.3390/molecules19033628.

[20] Carvajal-Millan E, Rascón-Chu A, Márquez-Escalante JA, Micard V, Ponce de León N, Gardea A. Maize bran gum: Extraction, characterization and functional properties. Carbohydrate Polymers. 2007;69(2):280-285. DOI: 10.1016/j.carbpol.2006.10.006.

[21] Niño-Medina G, Carvajal-Millan E, Lizardi J, Rascón-Chu A, Gardea A. Feruloylated arabinoxylans recovered from low-value maize by-products In: Jones CE, editor. Encyclopedia of Polymer Research: Nova Science Publishers, Inc.; 2011. p. 1401-1416.

[22] Iravani S, Fitchett CS, Georget DMR. Physical characterization of arabinoxylan powder and its hydrogel containing a methyl xanthine. Carbohydrate Polymers. 2011;85(1):201-207. DOI: 10.1016/j.carbpol.2011.02.017.

[23] Sárossy Z, Tenkanen M, Pitkänen L, Bjerre A-B, Plackett D. Extraction and chemical characterization of rye arabinoxylan and the effect of β-glucan on the mechanical and barrier properties of cast arabinoxylan films. Food Hydrocolloids. 2013;30(1):206-216. DOI: 10.1016/j.foodhyd.2012.05.022.

[24] Kačuráková M, Wellner N, Ebringerová A, Hromádková Z, Wilson RH, Belton PS. Characterisation of xylan-type polysaccharides and associated cell wall components by FT-IR and FT-Raman spectroscopies. Food Hydrocolloids. 1999;13:35-41.

[25] Egüés I, Stepan AM, Eceiza A, Toriz G, Gatenholm P, Labidi J. Corncob arabinoxylan for new materials. Carbohydrate Polymers. 2014;102:12-20. DOI: 10.1016/j.carbpol.2013.11.011.

[26] Mussatto SI, Dragone G, Roberto IC. Ferulic and p-coumaric acids extraction by alkaline hydrolysis of brewer's spent grain. Industrial Crops and Products. 2007;25(2): 231-237. DOI: 10.1016/j.indcrop.2006.11.001.

[27] Berlanga-Reyes CM. Carvajal-Millán E, Caire Juvera G, Rascón-Chu A, Marquez-Escalante JA, Martínez-López AL. Laccase induced maize bran arabinoxylan gels: structural and rheological properties. Food Science and Biotechnology. 2009;18:1027-1029.

[28] Martínez-López AL, Carvajal-Millan E, Lizardi-Mendoza J, López-Franco YL, Rascón-Chu A, Salas-Muñoz E, Barron C, Micard V. The peroxidase/H_2O_2 system as a free radical-generating agent for gelling maize bran arabinoxylans: rheological and structural properties. Molecules. 2011;16(12):8410-8418. DOI: 10.3390/molecules16108410.

[29] Ross-Murphy SB. Rheological characterisation of gels. Journal of Texture Studies. 1994;26:391-400.

[30] Ayala-Soto FE, Serna-Saldívar SO, Pérez-Carrillo E, García-Lara S. Relationship between hydroxycinnamic profile with gelation capacity and rheological properties of arabinoxylans extracted from different maize fiber sources. Food Hydrocolloids. 2014;39:280-285. DOI: 10.1016/j.foodhyd.2014.01.017.

[31] Berlanga-Reyes C, Carvajal-Millan E, Niño-Medina G, Rascón-Chu A, Ramírez-Wong B, Magaña-Barajas E. Low-value maize and wheat by-products as a source of ferulated arabinoxylans. In: Einschlag, editor. Croatia: InTech; 2011.

[32] Paes G, Chabbert B. Characterization of arabinoxylan/cellulose nanocrystals gels to investigate fluorescent probes mobility in bio inspired models of plant secondary cell wall. Biomacromolecules. 2012;13:206-214.

[33] Marquez-Escalante J, Carvajal-Millan E, López-Franco Y, Lizardi-Mendoza J, Valenzuela-Soto E, Rascón-Chu A, Faulds. Gels of water extractable arabinoxylans from a bread wheat variety: swelling and microstructure. Plants and Crop - The Biology and Biotechnology Research. iConcept Press. Retrieved from http://www.iconceptpress.com/books/plants-and-crop--the-biology-and-biotechnology-research/BK056-0101/.

Chapter 5

Perspectives on Biological Treatment of Sanitary Landfill Leachate

Abstract

Landfilling, one of the prevailing worldwide waste management strategies, is presented together with its benefits and environmental risks. Aside from biogas, another non-avoidable product of landfilling is landfill leachate, which usually contains a variety of potentially hazardous inorganic and organic compounds. It can be treated by different physico-chemical and biological methods and their combinations. The composition and characteristics of landfill leachate are presented from the aspect of biotreatability. The treatment with activated sludge, mainly consisting of bacterial cultures under aerobic and anaerobic conditions in various reactor systems, is explained, including an extensive literature review. The potential of fungi and their extracellular enzymes for treatment of municipal landfill leachates is also presented, with a detailed review of the landfill leachate treatment studies. The future perspectives of biological treatment are also discussed.

Keywords: Activated sludge treatment, biotreatability, fungal treatment, landfill leachate

1. Introduction

Landfilling is still widely accepted and used in any waste management strategy, but it can constitute a hazard for the environment. This method generally offers lower cost of operation and maintenance when compared to other methods, such as incineration. Besides households and urban activities, the industry is directly associated with the production of large amounts of solid wastes. Several methodologies and strategies have been developed for the integrated management of these wastes. They start with pollution prevention, waste minimization (*zero waste*), reuse of products or their parts, as well as material and/or energy recovery. But in spite of all environmental policies, the majority of municipal and industrial wastes still end up at the landfill and the amount of deposited wastes is significant worldwide. Landfill still accounted for nearly 40% of municipal waste treated in the European Union in 2010. In the 25 countries of the European Union, 502 kg of municipal waste was generated per person in 2010, while 486 kg of municipal waste was treated per person: 38% was landfilled, 22% incinerated, 25% recycled, and 15% composted.

In the deposited wastes, organics are still present even after thorough waste separation, mainly due to the dirty packages and other remains that could not be completely separated; thus, microbial processes dominate the stabilization of the waste and lead to the generation of the landfill gas, and dictate the amount and composition of the leachate. Landfill leachate is defined as wastewater formed due to precipitation, deposited waste moisture, and water, formed within the body of the landfill. Untreated leachates can permeate groundwater or mix with surface waters and contribute to the pollution of soil, ground water, and surface waters. Careful site management can reduce the quantity and increase the purity of the formed leachate, but it cannot completely eliminate it. Its composition is therefore site- and time-specific, based on the characteristics of deposited solid wastes, physico-chemical conditions, rainfall regime that regulates moisture level, and landfill age. Even within a single landfill site, variability is frequently evident [1, 2, 3]. Significant components of leachate at the beginning of landfill operation are heavy metals and degradable organics, while persistent organic pollutants usually appear later as a result of biotic (i.e., living components that constitute an ecosystem) and abiotic (i.e., non-living chemical and physical components that affect living organisms and the performance of ecosystems) processes in the system. Among these substances are several compounds classified as potentially hazardous: bio-accumulative, toxic, genotoxic (chemical compounds that damage the genetic information within a cell causing mutation that may lead to cancer), and they could have endocrine disruptive effect [2]. Hazardous substances from the leachate should be caught and removed properly, to avoid spreading in the receiving environment. Efficient treatment methods must be matched to the actual characteristics of a particular leachate and they could vary with time. Often, biological processes are employed if biotreatability in terms of low toxicity and at least moderate biodegradability of the leachate is indicated [2, 4].

Biodegradability of the wastewaters and also leachates is usually determined using various non-standardized laboratory or pilot-scale long-term tests with activated sludge as the source of active microorganisms [5]. Toxicity tests must be accomplished prior to the biodegradability determination to assess the impact of landfill leachate components on microorganisms of the aerobic or anaerobic activated sludge. Biodegradability assessment of leachates usually starts with the determination of ready biodegradation in common environmental conditions, it is upgraded with the assessment of biodegradation potential in an inherent biodegradability assessment test under optimal conditions, and it is finally concluded with a simulation of biodegradation in the wastewater treatment plant. All of the mentioned tests are based on the measurement of summary parameters, such as chemical oxygen demand (COD) or dissolved organic carbon (DOC) removal, O_2 consumption, etc. Inherent biodegradability assessment tests provide data on adsorption potential of the sample to the activated sludge and allow estimation of its impact on the biological wastewater treatment plant. Preliminary estimations should then be verified in a laboratory or a pilot-scale aerobic treatment plant to determine the actual impact of the wastewater on the activated sludge processes [5-7]. The connection between the biodegradation and changes in toxicity of the sample represents the stabilization study, where leachate is diluted in a batch reactor to avoid significant toxicity, and it is aerated and stirred until the biodegradation reaches the final plateau. Among other parameters, toxicity is monitored during biodegradation by using the appropriate method. Stabilization (ageing) allows us to assess the toxic fraction as permanent or biodegradable [8-10].

After the complete examination of the landfill leachate characteristics, the appropriate treatment process should be considered. In the case of significant biodegradability, various biological processes could be involved. Biological treatment is reliable, simple, highly cost-effective, and provides many advantages in terms of biodegradable and nitrogen and phosphorous compounds removal [11, 12]. It can be accomplished with microorganisms in different types of reactors, in aerobic and anaerobic conditions. Classical systems with activated sludge, sequencing batch reactor (SBR), biofilters, membrane bioreactors, as well as up-flow anaerobic sludge blanket processes and fluidized bed reactors are often considered [13]. In the last few decades, researchers also confirmed the great potential of white rot fungi for removal of hazardous as well as toxic pollutants. They produce various extracellular ligninolytic enzymes, including laccase (Lac) and manganese peroxidase (MnP), which are involved in the degradation of lignin in their natural lignocellulosic substrates [14], and according to the literature data [15], offer also an interesting potential for the landfill leachate treatment.

A considerable amount of work has been done in the field of landfill leachate biological treatment in the past decades. But the strict implementation of environmental legislative demands and the ageing of existing landfills put pressure on managers and operators of landfills to implement more efficient processes for landfill leachate treatment. Nevertheless, the results, obtained during decades of research, indicate some future research guidelines.

2. Solid waste management

The industrial and economic growth of many countries around the world has resulted in a rapid increase in industrial and municipal solid waste generation [1]. Many countries are still trying to implement separate collection and disposal of waste based on the Reduce-Recycle-Reuse principle, which should lead to zero waste management practice in the future. However, the improved waste management strategy, based on the waste management hierarchy pyramid (Figure 1) clearly pointed out the disposal of solid wastes, such as landfilling, as the least favorable option leading to severe environmental impact and degradation. It also results in a loss of natural resources [16]. Different recovery options, such as energy recovery, result in the utilization of energy potential of the waste, while recycling is even preferred due to the recovery of materials. In both cases, some loss of material and/or energy is noticed, and thus the differences regarding the impacts on the environment are relatively small [17]. The measures of waste prevention eliminate the need for the above-mentioned measures, thus lowering the impact that the waste has on the environment. The waste prevention measures can be any form of reducing the quantities of materials used in a process, or any form of reducing the quantity of harmful materials that may be contained in a product. At the same time, the replacement of hazardous materials should also be considered. Prevention and minimization also include processes or activities that avoid, reduce, or eliminate the waste at its source, or result in its reuse or recycling [18]. In recent years, these goals were incorporated in European environmental policy to make a resource-efficient Europe, while cutting off the burden of wastes and emissions. Effective implementation of these waste policies demands an understanding of what has been achieved so far and a set-up of targets and broader goals for the future. The first EU legislation covering the generation and treatment of waste was introduced in 1975, and now, there are more than 20 legislative documents currently in force on waste management [19-22].

Municipal solid waste is defined differently in different European states, but according to Eurostat, "Municipal waste is mainly produced by households, though similar wastes from sources such as commerce, offices, and public institutions are included. The amount of municipal waste generated consists of waste collected by or on behalf of municipal authorities and disposed of through the waste management system" [23]. Other wastes, similar in nature and composition, but collected by the private sector, are also included in this definition. In the EU Landfill Directive, municipal solid waste is defined broader as the "Waste from households, as well as other waste which, because of its nature or composition, is similar to waste from households" [24]. This seems to be the definitively more appropriate definition, referring to the type of the waste and not to who produced or collected it [25].

However, due to different definitions and methodologies used across EU countries, the efficiency of improved waste prevention in the last decade (2001-2010) is not easy to estimate. The average EU-27 value still varies around 500 kg for wastes generated per capita. 21 countries generated more municipal wastes per capita in 2010 than in 2001, and 11 cut per capita municipal waste generation [25]. On the other hand, there was a clear shift up the waste hierarchy, from landfilling to energy recovery and recycling. In this period of time, landfilling decreased by almost 41 million tons, whereas incineration increased by 15 million tons, and recycling grew by 28 million tons within EU-27. In the study where, together with the EU-27 countries, Croatia, Iceland, Norway, Switzerland, and Turkey were also taken into account, the decrease of landfilling in 2001-2010 is evident in Table 1 [25]. Each country can be included in several waste management categories, so the total number of countries is sometimes more than 32. When Table 1 is considered, the shift in recycling is noticeable; however, in 2010, in 50% of the countries landfill still presented more than 50% of municipal waste. The data on the trends in the recycling of materials and biowaste could be used to evaluate changes in the composition of deposited solid wastes. This parameter is also affected by the implementation of waste hierarchy, not only by the quantity of the deposited wastes. The total recycling rate increased between 2001 and 2010, mainly due to the fact that many countries have increased recycling of materials such as glass, paper and cardboard, metals, plastic, and textiles. Eight out of 32 investigated countries increased their material recycling rate by more than 10%, and 11 countries achieved an increase of 5%-10% [25].

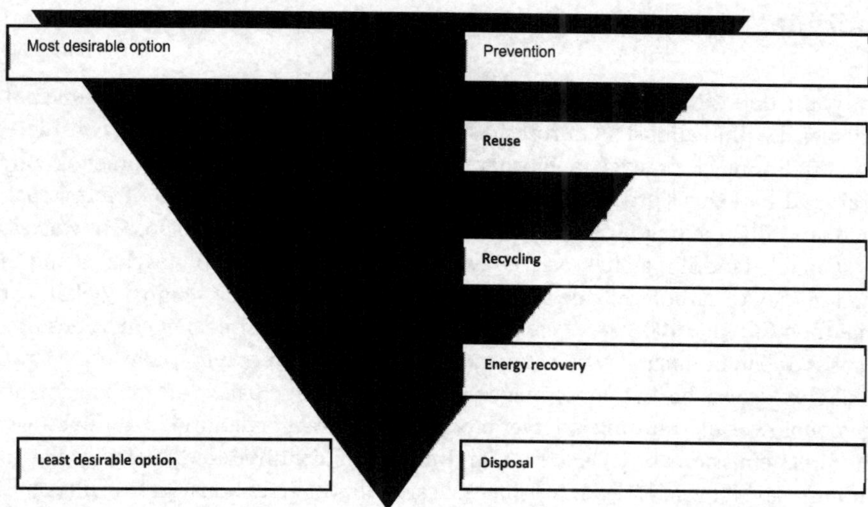

Figure 1. Waste management hierarchy [18].

Waste management	Number of countries	
	Year 2001	Year 2010
> 25 % Recycling	11	16
> 25 % Incineration	8	10
> 50 % Landfilling	22	19
> 75 % Landfilling	17	11

Table 1. Number of countries at different levels of municipal waste management in 2001 and 2010 [25].

In contrast to material recycling, biowaste recycling is not comparably efficient over the same period of time (EEA, 2013); one country increased its municipal waste-derived biowaste recycling by more than 10%, and only six of them improved it by 5%-10%. Eighteen countries even sustained a very low level of biowaste recycling (0%-10% of municipal waste generated). This could be a consequence of several factors affected by the methodology of data collection and the legislation. The fact is that material and biowaste recycling potential depends on the share of each waste type in the total collected municipal waste. In most countries, the biowaste recycling potential is lower than the material recycling potential because biowaste represents a smaller proportion of the total municipal waste [25]. This leads to the conclusion that different policy instruments should be implemented and combined in different countries to achieve maximal impact in terms of municipal waste management.

3. Landfilling of municipal waste

Currently, the deposit in a landfill is still the most widely used method for municipal solid waste disposal within almost all European states (Table 1). The landfill can be considered as a complex environment or even a biochemical reactor, where many interacting physical, chemical, and biological processes take place. The degradation process of municipal solid wastes in landfills is a long-term event. A major problem regarding disposal of wastes is the lack of available landfilling sites, as well as the production of landfill leachates and biogas, consisted mainly of carbon dioxide and methane, which has 28 times higher global warming potential than CO_2 in a 100-year cycle [26]. In Slovenia, for example, net emissions of greenhouse gases due to municipal waste management have been decreasing constantly since 1999 [27]. This could also be the consequence of a better municipal waste management (e.g., recycling of biowaste), resulting in lower biodegradable fraction landfilled. It can be assumed that the direct emissions of GHCs from landfilling will continue to decrease in the coming years, but for several years ahead, considerable amounts of greenhouse gases will continue to be emitted from landfills because biowaste landfilled in the previous years will continue to generate methane for several decades [27]. However, with appropriate entrapment and utilization, biogas is usually efficiently exploited for energy purposes at the site, while leachates could pose a serious risk for nearby soil, surface, and underground waters [28]. At the sites, where there is no need for energy or where the methane content is very low, methane is flared to avoid its migration in the atmosphere. Landfill top covers or so-called biocovers are often used at landfills to reduce methane emissions. They optimize environmental conditions for development of methanotrophic bacteria to help oxidize any fugitive methane. Biocovers are usually spread over the entire surface of the landfill and they are made of compost, dewatered sewage sludge, or other waste material. Landfills with gas collection and recovery systems had a methane recovery efficiency of 41%-81%. Methane emissions could range from 2.6 kg h^{-1} to 60.8 kg h^{-1}, with the lowest emissions from the small and old landfills and the highest emissions from the larger landfills [29].

3.1. Landfill leachate

Appropriate management can reduce the quantity and quality of the leachate, but it cannot be completely eliminated. Landfill leachate is generated as a mixture of rainwater percolation through the wastes, water produced from the (bio) degradation of wastes, and the water present in the wastes at the time of deposition [2]. Its composition is based on the type and the amount of waste deposited, as well as its maturity, and the construction of the landfill site [4] Main sequential and distinct consecutive stages involved in landfill stabilization are [30, 31]:

1. Aerobic phase, characterized by processes enabled by oxygen present. In this acclimation phase, sufficient moisture develops to support active microbial communities. Initial

changes in compounds occur in order to establish the appropriate conditions for further biochemical degradation.

2. Hydrolysis and fermentation stage, where complex molecules are broken into smaller fragments and an aerobic environment is transferred to an anaerobic one, the amount of entrapped oxygen is drastically reduced and the reducing conditions occur. The main electron acceptors are nitrates and sulfates.

3. Anaerobic acetogenic stage, characterized by continuous hydrolysis of solid wastes, preceded by the production of volatile fatty acids at high concentrations. Low hydrogen levels promote the activity of methanogenic bacteria, which produce methane and carbon dioxide from organic acids. This stage could be recognized by a high concentration of metals in the leachate, due to their increased mobility because of the lower pH, even below 4.

4. Anaerobic methanogenic stage, where significant methane production is evident. pH is again increased to 7-8, due to the degradation of intermediate acids and the buffering capacity of bicarbonates. The concentration of heavy metals is reduced again, due to their complexation and precipitation. During this stage, the mesophilic bacteria, which are active in temperatures of 30°C-35°C, and thermophilic bacteria, active in the range of 45°C-65°C, dominate the microbial population.

5. Maturation stage, as the final stage of landfill stabilization, could be recognized by low microbial activity due to degradation of biodegradable fraction and limiting impact of nutrients. As a result, the methane production decreases, as does the amount of pollutants in the leachate, which usually stays at a constant level. Slow degradation activity of the resistant organic pollutants can be observed by the production of humic and fulvic substances.

The duration of each phase is dependent upon many factors, and the development and activity of microorganisms are dependent upon sustaining appropriate conditions. Landfill leachates are characterized by high concentrations of numerous toxic and carcinogenic chemicals, heavy metals, and organic as well as inorganic matter. Among the organic compounds detected in the landfill leachate, the main compounds are different hydrocarbons, esters, alcohols, and ketones, as well as aromatic and heterocyclic compounds [32]. Additionally, the leachates can also be contaminated with bacteria, including aerobic, psychrophilic and mesophilic bacteria, coliform and fecal coliforms, spore-forming-bacteria, and with numerous fungi [32, 33]. Typical concentrations of landfill leachate compounds as a function of landfill age and stabilization are presented in Table 2 [30, 31, 34].

Concentrations of organic compounds, expressed as COD and biological oxygen demand 5-day test (BOD_5), in the young leachate are high (aerobic and acid formation phase), while leachates from stabilized landfills (methane formation and maturation phase) contain lower levels of organic matter [28]. Several authors have reported that in young landfills, COD could vary from 5,000 to even more then 60,000 mg L^{-1}, with BOD_5 starting from 3,000 L^{-1} to even 40,000 mg L^{-1} [35]. Those values are significantly lower in leachates from mature landfills, as a result of biotic stabilization processes in the body of the landfill [36]. The total quantity of the produced landfill leachate can be estimated by using empirical data based on flow measurements, or by using water mass balance between precipitation, evapotranspiration, surface runoff, and capacity of moisture storage. Waterproof covers and different covering liners contribute a lot to the reduction of landfill leachate quality, but they cannot completely reduce it. Both parameters (leachate quality and quantity) affect the attempts of uniform design of leachate treatment systems. The optimal treatment solution may change over time because of the changeable quality and quantity of the leachate, and the development of new technologies and the legislation [37].

Parameter	Aerobic phase	Acid formation phase	Methane formation phase	Maturation phase
pH (/)	6.7	4.7–7.7	6.3–8.8	7.1–8.8
COD (mg L^{-1})	480–18,000	1500–71,000	500–10,000	<1000
BOD_5/COD (/)	>0.5	0.5–1.0	0.5–1.0	<0.1
Organic compounds (/)	Various	80% of volatile fatty acids	5%–30% of volatile fatty acids + fulvic and humic acids	fulvic and humic acids
Ammonium nitrogen (mg L^{-1})	>100	<1000	<500	<500
Heavy metals concentration (/)	Low	Low - Medium	Low	Low
Conductivity (μS cm^{-1})	2,500–3,300	1,600–17,100	2,900–7,700	1,400–4,500
Biodegradability	Important	Important	Medium	Low

Table 2. Typical concentrations of landfill leachate concentrations as a function of landfill stabilization [30, 31, 34].

3.2. Landfill leachate treatment

The landfill treatment and disposal represents one of the major landfill operational costs. Its management is not accomplished by the closure of the landfill, because its characteristics must be monitored even further, as defined by particular legislative requirements.

Traditionally, biological treatment is the most widely-used treatment strategy for wastewaters, mostly because of its low operational costs and complete as well as rapid destruction of pollution [2]. However, biological treatment is not always effective enough for the toxic and recalcitrant leachates (e.g., methane formation and maturation phase), in which case physico-chemical treatment can take place. Usually, the leachate treatment involves a combination of various biological and chemical methods. Conventional landfill leachate treatment can be classified into three major groups:

1. Recycling and combined treatment with domestic sewage. In the past, it was common to treat landfill leachate mixed with municipal wastewater [2, 38]. This option is not so favorable nowadays, due to the identified presence of hazardous persistent compounds in the leachate, which are not removed in the conventional municipal treatment plant. On

the other hand, the municipal wastewater represents an important source of nitrogen and phosphorous, which would otherwise be limiting factors in the biological treatment of landfill leachate alone [39]. The new trends include recycling of landfill leachates, to manage the landfill as a bioreactor with moisture and air control to enhance the establishment of conditions for efficient biodegradation of present organic fractions. However, this can decrease the concentration of organic constituents in the landfill leachate, but it can also increase the concentration of ammonium nitrogen, which should then be removed by additional treatment processes. Recycling of the leachate is a viable option especially in developing countries, to reduce its environmental risks and to avoid as many multi-process treatment methods as possible [40]. Another important aspect is also the reduction of time needed to stabilize the deposited waste, from several decades to a few years [41].

2. Biological treatment employing mainly aerobic biodegradation processes. Biological treatment is reliable, simple, highly cost-effective, and provides many advantages in terms of biodegradable matter and nitrogen compounds removal. However, its efficiency is strongly limited in the presence of refractory or inhibitory compounds in wastewaters, which are also typical for mature landfill leachates. All of the aspects of the biological treatment of applications will be presented in the next chapters.

3. Chemical and physical methods, such as chemical oxidation, adsorption, coagulation/flocculation, membrane techniques and air stripping [2]. Membrane-based treatment processes gained a lot of attention recently, including reverse osmosis, nanofiltration, ultrafiltration, and microfiltration [2, 31]. Chemical oxidation processes are viable options for effective landfill leachate treatment, but they are relatively expensive for the complete mineralization of landfill leachate pollutants, because the oxidation intermediates, formed during treatment, tend to be more and more resistant to the complete chemical degradation. Fenton oxidation or ozonation in a pre-treatment process can convert persistent and non-biodegradable organic compounds into more biodegradable intermediates, which would be subsequently treated by a biological treatment process. On the other hand, the polishing of remained organics in the effluent after biological treatment using one of the oxidation processes could also reduce the environmental impact of landfill leachates. Several authors have reported that Fenton's process can achieve 60%-90% of COD removal of organics from landfill leachate [42]. Moreover, if leachates were pretreated by biological processes, the Fenton's oxidation added additional 63% of COD removal [43]. Some authors also suggested electrocoagulation as a suitable method for leachate treatment, accompanied by a flotation of formed sludge [44]. In this study, 40-73% of COD and color at 245 nm were removed, depending upon different cathode material (graphite or aluminum). However, the treatment lasted for 210 minutes and the energy requirement per kg of removed COD was 135 kWh, accompanied by 40% and 0% of nitrate and ammonium nitrogen removal, respectively.

3.3. Selection of the treatment process

It is well recognized that the selection of appropriate treatment method is strictly dependent upon the leachate characteristics and composition. For effective biological treatment, its biotreatability must be evaluated. In the case of a low-treatment efficiency of the biological plant, other treatment methods should be investigated.

Any determination of biological treatability of the wastewater must include data on toxicity and biodegradability. Toxicity could be assessed using one of the tests with mixed culture of microorganisms (activated sludge), which plays an essential role in a biological wastewater treatment plant. The test with measurement of inhibition of oxygen consumption and the test where growth inhibition of activated sludge is measured are widely applied [45]. If the impact of the leachate to anaerobic microorganisms is assessed, then the reduction of biogas production is often measured [46].

Method	**Group of pollutants removed**	**Data obtained**
Filtration at different pHs	Suspended solids	· Toxicity is related to soluble or insoluble material · Metals form insoluble complexes at higher pHs and are removed during filtration
Ion exchange	Inorganic compounds ions	· Toxicity is related to inorganic compounds or ions
Biodegradability testing	Biodegradable fraction	· Decrease in toxicity due to the biological treatment · Toxicity is related to recalcitrant or biodegradable compounds · The extent of biodegradation of wastewater at investigated condition · Possible sorption of pollutants to microorganisms. · Biodegradability or mineralization potential
Oxidant reduction	Oxidants	· Toxicity is related to oxidants.
Metal chelation (EDTA)	Cationic metals (no Hg)	· Toxicity is related to metals.
Air stripping at different pHs	Ammonia Volatile organics	· Toxicity is related to volatile organics · At low pH (pH = 3) small molecular weight organic acids will be effectively removed, while at higher pH (pH = 11) they may be dissociated or form salts and are not purged out · Toxicity is related to ammonia · Ammonia could be stripped out at higher pH.
Adsorption	Adsorbable organics	· Toxicity is related to adsorbable organics · Presence of compounds causing color · Results very dependent upon adsorbent used and experimental conditions
Oxidation	Oxidizable organics	· Toxicity is related to oxidizable organics · Results very dependent upon oxidant used and experimental conditions

Table 3. Commonly used methods in Toxicity Identification Evaluation (TIE) procedures [12, 49, 57].

Biodegradability of the wastewaters is usually determined by using various non-standardized laboratory or pilot-scale long-term tests with activated sludge as a source of active microorganisms [5, 6]. At the same time, some of the standardized test methods, developed for biodegradability assessment of pure chemicals, could be applied [7]. Biodegradability assessment usually starts with the determination of ready biodegradation in common environmental conditions and it is upgraded with the assessment of biodegradation potential in the inherent biodegradability assessment test under optimal conditions [5]. All of the above mentioned tests are based on the measurement of summary parameters, such as COD or DOC removal, O_2 consumption, etc. [47, 48]. A combination of different measurement techniques to follow biodegradation is recommended to distinguish between the biodegradation and the complete mineralization of the sample [5]. Inherent biodegradability assessment tests provide the data on adsorption potential of the sample to the activated sludge and allow us to estimate its impact on biological wastewater treatment plants [49]. Preliminary estimation should then be verified in an actual laboratory or a pilot-scale aerobic treatment plant, to determine the impact of the wastewater on activated sludge processes. Another possibility to assess biotreatability is also the stabilization study, which represents a link between toxicity and biodegradability, to correlate the changes of toxicity versus the extent and rate of the biodegradation. The initial and final biodegradability testing of the test mixtures allows us to confirm the measured degradation or the persistency of final residue [10]. If the wastewater consists of mainly degradable components, resulting in the toxicity elimination, the biological treatment is a good alternative, while in the case of poorly biodegradable wastewater with negligible decrease in toxicity, other treatment methods should be considered [8-10].

However, the discussion on the selection of the treatment method is based on the knowledge on wastewater quantity and quality, as well as the required effluent quality. The costs and the availability of the land are also very important; a detailed cost analysis should therefore always be made prior to the final process selection and design. The main characteristics, which should be considered are [12]: i) soluble organics responsible for oxygen consumption; ii) suspended solids; iii) priority substances that have hazardous environmental impact due to their persistency, toxicity, bioaccumulation potential and they could pose endocrine disruptive effect; iv) heavy metals; v) substances and particles causing color and turbidity; vi) nitrogen and phosphorous content; vii) refractory substances; viii) floating oils and grease; ix) volatile compounds (organics and H_2S), etc. For wastewaters containing nontoxic and biodegradable organics, the process design criteria can be obtained from the data from laboratory or pilot studies, while more defined screening procedures are often needed for more complex and changeable wastewaters, such as landfill leachates. To set up appropriate treatment technology, the toxicity identification (TIE) approach is sometimes feasible, especially when a biological treatment is considered [53]. The TIE is a wastewater-specific study to isolate, identify and confirm the causative agents of toxicity. It is based on procedures, developed by the United States Environmental Protection Agency (USEPA). The Toxicity Reduction Evaluation (TRE) procedure is used as a tool to identify toxic components that may be removed or reduced in an effluent to reduce toxicity problems.

Combined processes	Experimental scale	Type of the leachate	Measured parameters	Removal efficiency	Reference
· Sequencing batch reactor (SBR) · Coagulation with polyferric sulphate + Fenton system · Upflow biological aerated filters	Full	Mature/ Stabilized	COD NH_3-N TP SS	97.3% > 99% < 1 mg/L-1 < 10 mg/L-1	[62]
· Sequencing Batch Biofilter Granular Reactor · With/no ozone · Solar photo-Fenton	Laboratory	Medium aged	Toxicity: -respirometry -*Vibrio fischeri* -*Lepidium sativum* COD DOC	 High High High 95.3% 95%	[63]
· $FeCl_3$ coagulation · Magnetic ion exchange · Reverse osmosis · Nanofiltration	Laboratory	Mature/ Stabilized	DOC UV245 adsorbing OM Salts DOM	45–71%* 84–94%* > 93%* > 99%*	[64]
· Aerated lagoon · Solar photo-Fenton · Conventional biological WWTP (with nitrification/ denitrification)	Pilot	Young/ After lagooning	DOC TN Biodegradability	90% 56%–90%* Increased	[61]
· Agitation/stripping · $FeSO_4$ coagulation · SBR, mixed with sewage: anoxic-aerobic-anoxic conditions · Sand and carbon filtration	Laboratory	Mature/ Stabilized	COD TOC BOD_5 SS NH_3-N	97.4% 92.3% 94.4% 97.5% 99.2%	[66]
· TiO_2/UV photolysis · Bioreactors with various inoculums (raw leachate/soil extract/activated sludge)	Laboratory	Mature/ Stabilized	COD BOD_5 NH_3-N	87% 90% 43%–79%*	[65]
· Coagulation/flocculation · Fenton · Solar photo-Fenton	Laboratory Pilot	Mature/ Stabilized	Toxicity: -respirometry COD Biodegradability	 Remains low 89% Increased	[67]

*...Depends upon the combination of treatment processes and landfill leachate sample characteristics.

TP = Total phosphorous

TN = Total nitrogen

SS = Suspended solids

COD = Chemical Oxygen Demand

DOC = Dissolved Organic Carbon

OM = Organic Matter

DOM = Dissolved Organic Matter

WWTP = wastewater treatment plant

Table 4. Some of the recently investigated combinations for treatment of heavily polluted landfill leachates [61-67].

The TIE methodology uses the responses of the test organisms to detect the presence of toxic substances in the sample before and after the samples are subjected to a series of physical and chemical treatments. This combination of physical/chemical manipulations of toxic samples, followed by the toxicity testing, allows one to isolate and identify the problematic group of compounds [50, 54, 55]. The most often used procedures are listed in Table 3. Results could be efficiently utilized to set up the appropriate treatment procedure for the particular wastewater, because it can be clearly estimated which treatment method is efficient in the removal of the particular group of pollutants and where the toxicity of the wastewater comes from [56]. Usually these simple, cost-effective methods could reduce the need for a complex and detailed characterization of the wastewater before setting up treatment procedure. However, they have to be designed and performed with caution (blank sample) to avoid any impact of the applied method (pH manipulation, addition of chemicals, etc.) to the final characteristics of the sample.

According to the results of the methods described in Table 3, suitable treatment methodology could be set up. If, for example, the wastewater contains a significant fraction of a non-biodegradable organic fraction (determined by biodegradability testing) and it contains a lot of oxidizable organics (proved by oxidation experiment), one of the advanced oxidation processes would seem to be to be the most viable treatment option. On the other hand, if it contains organics that are able to mineralize almost completely in the biodegradability test, and it contains a lot of ammonia, one of the biological treatments involving nitrification/denitrification would seem to be to be the best choice. It can be clearly concluded that, in the case of municipal landfill leachates, a technically and economically viable methodology for the effective treatment has yet to be designed. The available options are similar to those used in the treatment of industrial wastewaters, involving a combination of physical, chemical, and biological processes. Primarily due to their low costs, the biological processes, in their various forms according to redox regime (aerobic, anaerobic, anoxic), a type of biomass (a mixed or a pure bacterial culture, fungi, etc.) and a biomass fixation (dispersed, attached), remain the most widely implemented type of treatment processes [2, 58, 59].

However, a combination of biological and physico-chemical processes is usually employed for heavily polluted leachates. Many examples of efficient treatment combinations could be found in literature. As presented in [60], an aerobic biological treatment, a chemical coagulation, an advanced oxidation process (AOP), and some combined treatment strategies were compared. Laboratory experiments were done with 200 mL samples in a glass vessel. The efficiency of these treatment procedures was evaluated by analyzing the COD and color removals. In the extended aeration process, the maximum COD and color removals were 36% and 20%, respectively. They could be achieved during the optimum retention time of 7 days. Chemical coagulation with an optimum aluminum sulphate dose of 15,000 mg/L at pH = 7.0, gave the maximum COD and color removals of 34% and 66%, respectively. Using Fenton oxidation process at optimum pH = 5.0 and optimum dosages of reagents, with H_2O_2/Fe^{2+} molar ratio of 1:3, the highest removals of COD and color were 68% and 87%, respectively. The combined treatment, the extended aeration followed by Fenton oxidation, was found to be the most suitable.

Some additional, recently investigated and proposed treatment designs are presented in Table 4.

A large-scale multistage treatment system was also designed for the treatment of a mature raw landfill leachate [61]. The system consisted of an activated sludge biological oxidation (ASBO) reactor for aerobic and anoxic conditions (volume 3.3 m^3) and a solar compound parabolic collector (CPC) for photo-Fenton process (total collector surface 39.52 m^2 and illuminated volume 482 L). The raw leachate was characterized by a high concentration of humic substances, representing 39% of the DOC content and high nitrogen content, mostly in the form of ammonium nitrogen. In the first biological oxidation step, a 95% removal of total nitrogen and a 39% mineralization in terms of DOC were achieved. The following photo-Fenton reaction led to the depletion of humic substances > 80% of low-molecular-weight carboxylate anions > 70% and other organic micropollutants, thus resulting in a total biodegradability increase of > 70%. The neutralized photo-bio-treated leachate was finally treated with the second stage biological oxidation, where the rest of biodegradable organic carbon and nitrogen content were eliminated. This way, a high efficiency of the overall treatment process was achieved.

4. Biological treatment of the landfill leachate

4.1. Treatment with activated sludge

Biological treatment has become one of the most often used treatment processes; it is the most common method for the removal of organic, nitrogen, or phosphorus components from wastewaters. One of the main reasons for the selection of this process is its capability to achieve high elimination efficiency of these pollutants, and at the same time, it is relatively less expensive than physico-chemical or chemical processes. The pollution is completely destroyed to the level of non-hazardous, simple products, and not only transformed into another form. Nowadays, it is used not only for the treatment of sewage, but also for the removal of different xenobiotics such as pharmaceuticals, personal care products, and cleaning agents from the sewage and the heavily polluted industrial wastewaters and landfill leachates [12]. Biological degradation of pollutants is caused by the metabolic activity of microorganisms, in particular by the bacteria and fungi that live in natural environments. However, its efficiency is strongly reduced in the presence of refractory or inhibitory compounds in wastewaters, typical also for mature landfill leachates [33, 66, 68]. To achieve good removal efficiency, high BOD/COD ratio is recommended (>0.5) [13].

When biological treatment is discussed, mainly microorganisms that grow in a controlled environment through a complex sequence of biochemical reactions, forming the vital steps of their metabolic activities are considered [13]. The prevailing species are the saprotrophic bacteria, there is also an important protozoan flora present, composed mainly of amoebae, *Spirotrichs, Peritrichs* including Vorticellids, and a range of other filter-feeding species. Fungi could also contribute to the diversity of present populations. Other important constituents include motile and sedentary rotifers. The most important seems to be the bacteria, found in all types of treatment processes. The nature of the population changes continually, in response to variations in the composition of the wastewaters and to environmental conditions [69]. Generally, biological treatment of wastewaters involves bacterial community in aerobic and anaerobic conditions, which could be dispersed or attached, in small flocks, granulated or forming biofilms, while treatment with fungi and their enzymes also lately received more attention [70]. However, fungal treatments have not yet found a wider recognition, due to the difficulty in selecting organisms that are able to grow and remain active in the actual wastewater [71].

4.1.1. Treatment with activated sludge under aerobic conditions

The most often applied processes for biological treatment are aerobic. In an aerobic environment (concentration of dissolved oxygen >2 mg L^{-1}), organic matter is used as a food source for microorganisms. The suspended organics are removed by entrapment in the biological activated sludge flocks. The colloidal organics and a small amount of soluble organics are also partially adsorbed and entrapped by the sludge flocks. Therefore, approximately up to 85% removal of the total COD can be achieved after 10 min to 15 min of retention time. The remaining degradable soluble organic fraction undergoes biological reactions [11, 13]. A portion of organic compounds (about 50% of organic carbon) is oxidized to CO_2 and H_2O, and the rest of it is incorporated into a new biomass. Approximately about 60% of the energy content in wastewater organics is consumed for synthesis of the new biomass and the rest represents reaction heat loss [13]. At the same time, an efficient removal of ammonium nitrogen should also be achieved to protect the sensitive water bodies from eutrophication [12, 54].

A combination of aerobic and anoxic environment is necessary for the accomplishment of organic and nitrogen pollutants' removal from wastewater. On the other hand, the combination of anaerobic and aerobic environment is required to biologically remove phosphates from wastewater, so the systems could not always be characterized as completely aerobic or anaerobic. The upgrading of biological processes, from removal of carbonaceous organics to the nitrogen and phosphorus removal, significantly impacted the system configuration. Not only do the system configuration and its operation increased in complexity, but also the new legislative demands on effluent quality have to be met. Thus, the system must be well-designed, optimized, and operated at its optimum in order to fulfill these criteria [72]. Some of the bioreactors, also applicable for aerobic treatment using microorganisms, are summarized in Table 5. Aerobic biological systems, based on suspended-growth biomass, have been widely studied and also applied [2]. Also recently, attached-biomass systems have been developed, such as moving bed bioreactors (MBBR) and with different options of biofilters. A promising alternative also are membrane bioreactors (MBR), which represent an advanced biological treatment process, replacing a secondary clarifier in activated sludge process for removal of biomass with membrane module. It can be incorporated as an internal or an external unit of aeration basin to achieve better effluent quality, process stability, increased biomass retention time, and low sludge production [73]. Some of the systems where biological treatment represents the most effective stage will be briefly overviewed and discussed in this chapter. It should be emphasized that sometimes it is difficult to distinguish between aerobic and anaerobic treatment plants, due to the fact that most of the systems apply a combination of different regimes (aerobic, anaerobic, anoxic) to achieve the optimal treatment efficiency.

Systems with suspended-growth biomass	Systems with attached/immobilized biomass
Lagoons: · Aerated · Non-aerated	Biofilters: · Tricking filters · Submerged biological filters
Constructed wetlands: · Horizontal system · Vertical system	Moving bed bioreactors (MBBR): · Rotating biological contactors · Suspended carrier biofilm reactors
Activated sludge (AS) systems: · Continuous flow reactors · SBR	
Membrane bioreactors (MBR): · External membrane module · Submerged/Immersed membrane module	

Table 5. Typical treatment system with aerobic microorganisms applied in landfill leachate treatment [2, 13, 73].

When aerated and non-aerated lagoons are discussed, together with artificial or natural wetlands, there is usually a combination of aerobic-anaerobic systems. The upper part is usually aerated, while the bottom part is anaerobic [74]. Such combination is well-illustrated in [75]. Four connected on-site lagoons were used for the treatment of mature landfill leachate in the aging methanogenic state. The landfill leachate contained a relatively low COD value (mean value 1,740 mg L^{-1}) and a relatively high ammonium nitrogen concentration (mean value 1,241 mg L^{-1}). The pH of the raw leachate was in the range of 7.0-8.0 and the temperature was 16.7ºC, which is higher than the mean ambient temperature (13.5ºC). Volumes of the lagoons varied from 60-80 m^3. The leachate was mixed and aerated by compressed air (4-6 h) through diffuser pipes at the base of lagoons. The facultative aerobic system was obtained where sequential aerobic and anaerobic stages were maintained. The total COD removal was 75% in 56 days. Ammonium nitrogen removed 99%, while an average of 9 mg L^{-1} remained. However, the authors calculated that the conditions at the site could still be toxic to fish and the treated leachate should be diluted before its release into the environment. Due to aerobic conditions, nitrification occurred and the concentration of the nitrate was considerably higher in the effluent of the lagoons than in the influent; however, 80% of nitrogen was removed. The bacterial community profile also differed from one lagoon to another.

The study by [76] compared efficiency of horizontal- and vertical-constructed wetlands for landfill leachate, containing 2,930-14,650 mg L^{-1} of COD, 170-4,012 mg L^{-1} of ammonium nitrogen, and 44-153 mg L^{-1} of ortophosphate-P. The experiments were run in a continuous mode, in three subsurface wetland systems; two of them operated in a vertical flow mode and one in a horizontal flow mode. The system was planted by *Typha latifolia*. Basins were 100 cm in length, 50 cm wide and 40 cm deep. The systems were filled with different heights of gravel and sand when zeolite was added in the third lagoon to increase its adsorption and ion exchange capacity. The leachate was introduced intermittently (10 min h^{-1}) to assure hydraulic retention time of 8-12.5 days. In the vertical systems, COD removal was 15%-42%, while it reached up to 61% in the horizontal one. Ortophosphate-P removal in the vertical systems was 30%-83%, while in the horizontal one, it varied between 26.3 and 61.0%, depending on the climate conditions (a month of determination). The removal of NH_4-N was better in vertical systems (36.8-67.4) in comparison to the horizontal one (17.8-49.0). The authors also presented the removal of heavy metals from the leachate. The concentration of Cr and Zn increased in the effluent, suggesting that they were washed out of the system. The iron removal decreased with time and was below 50%. The removal of lead varied between 30%-90%. It was concluded that the vertical system with zeolite layer was beneficial, especially in terms of ammonia removal.

In the study of [77], the activated sludge system with sequencing batch laboratory reactor mode was employed for the treatment of landfill leachate, containing 4,298-5,547 mg L^{-1} of COD, 913-1,017 mg L^{-1} of BOD_5, 13,971-17,421 mg L^{-1} of TDS, and 72-374 mg L^{-1} of NH_4-N. Biomass was in the form of granules (0.36-0.60 mm, in cylindrical shape), the working volume of the reactor was 3 L in the operational mode of 12-h cycle (60 min of feeding, 640 min of aeration, 5 min of settling, and 5 min of effluent discharge). The reactor operated for three months with 182 cycles. The average COD removal was 82.8%-84.4%, while the ammonium removal efficiency reached 62±8%. The authors confirmed the high impact of initial ammonium N concentration on the performance of nitrification, which was confirmed by the simultaneous experiment with pretreated landfill leachate, where NH_4^+-N was reduced. To obtain a high removal of organics, a pretreatment in terms of ammonium removal was proposed.

Removal of nutrients, especially ammonium nitrogen, from landfill leachates using a biological system was also studied by [78]. A batch reactor (V=150 L) was used for the treatment of homogenous mixture of old and freshly produced leachate from the body of the Tunisian landfill. The first treatment step involved an anoxic process preceded by aerobic processes. In the anoxic phase, COD reduction reached 46%, TOC was reduced to 65%, while ammonium N removal was 45%. Afterwards, the treated leachate was led to three aerobic submerged biological reactors, where a 7-day retention time was employed. As a result, high overall treatment efficiencies were obtained in BOD_5, COD, and the NH_4^+-N removal was 95%, 94%, and 92%, respectively.

MBR also gained a lot of attention lately for the treatment of landfill leachate [73]. They are often referred to as an efficient and robust alternative to other systems, in spite of their higher operational costs. They are essentially composed of two parts; i) a bioreactor dealing with the removal of organics and ii) a membrane module for the separation of the treated leachate and biomass. In comparison to the conventional activated sludge systems, they allow for complete retention of the biomass in the system; this way, the settling characteristics of the sludge are less important. As a result, the system can be operated at much higher concentrations of biosolids, up to 20 mg L^{-1}, with very clean effluent. Additional important advantages are also higher loading rates, smaller volumes, lower production of excess sludge, and easier development of microorganisms with lower growth rates. MBR systems are usually designed as ultrafiltration or nanofiltration in a hollow fiber, plate, or frame; they could be in a flat or tubular configuration with continuous stirred tank reactors, plug-flow systems or sequencing batch reactors.

A submerged MBR for the treatment of heavily polluted landfill leachate, containing 18,685 mg L^{-1} of COD and 310 mg N per liter was used in reference [3]. Biologically treated leachate was additionally filtrated, using nanofiltration and reverse osmosis. The biological stage was very effective, removing 89% of COD and 85% of the total Kljeldahl nitrogen (TKN). In MBR, polyeter sulfone ultrafiltration membrane was installed as a submerged module. A system with the volume of 4 L operated as SBR. The landfill leachate was fed daily (300 mL) and the lack of phosphorus was overcome by the addition of KH_2PO_4. The system reached steady-state operational mode after 4 months with 9 g L^{-1} of biomass. COD removal stabilized at 89%. Some portion of inert COD was entrapped in the reactor. Due to the unstable pH conditions, effective nitrification was not achieved; it was probably also reduced due to the air stripping of NH_4-N at pH=8.6 and the consequent lack of the ammonium.

However, sometimes it is impossible to distinguish clearly between systems with suspended and attached biomass, since the combination of advantages of both treatment systems is sometimes the most optimal to assure the environmentally and legislatively acceptable performance. A combination of a cross-flow MBR and MBBR for the treatment of stabilized leachate was, for example, used in reference [79]. The treatment was focused on the removal of ammonium N, present in stabilized landfill leachate up to 3,000 mg L^{-1}. A combination of pure oxygen MBR and the subsequent MBBR was used for the nitrification and the denitrification, respectively. The volume of the membrane bioreactor was 500 L and it contained ultrafiltration ceramic membrane, while the volume of the MBBR was 540 L. Ammonium was oxidized only to a nitrite to reduce oxygen consumption for 25%, in comparison with the much higher oxygen consumption in the case of its complete conversion to nitrate. The authors obtained a 90% conversion of the ammonia N to nitrite with the sludge retention time of over 45 days. The system also enables up to 40% savings in the COD demand for nitrification. It was possible to oxidize more than 95% of total N inflow, and effluent ammonia concentrations were always below 50 mg L^{-1}.

Fixed bed filters offer a higher resistance to the toxic compounds and lower temperatures [80]. Aerated filters have been used for efficient treatment of landfill leachate [81]. In the study of [82], two systems using attached biomass were compared. Two MBBR systems using small free-floating polyurethane elements and granular activated carbon (GAC) as biomass carriers were set up. The laboratory SBR system had a volume of 6 L, and the inflow leachate contained NH_4-N, COD, and BOD_5 in the average values of 1,800 mg L^{-1}, 5,000 mg L^{-1}, and 1,000 mg L^{-1}, respectively. According to the low BOD_5/COD ratio=0.2 and pH=7.5, the landfill leachate could be characterized as a stabilized one. The study was conducted in two separate treatment cycles, the first one with cube-shaped polyurethane (30 g/reactor), while in the second one, 90 g of GAC (1,100 m^2 g^{-1}) was added. Nitrification and denitrification processes also occurred, but the need for additional external dosage of carbon source was emphasized. However, these processes efficiently and almost completely removed nitrogen, accompanied by the sufficient removal of COD (up to 81%), BOD_5 (up to 90%), and turbidity.

A bioreactor cascade with a submerged biofilm was also successfully used for a young landfill leachate treatment. Three reactors, each with the working volume of 18 L were used. The biofilm support was made from PVC synthetic fiber (57 m^2 m^{-3}). The reactors were inoculated with sludge from wastewater treatment plant and the biofilm consisted of various microorganisms, while bacterial groups such as *Bacillus, Actinomyces, Pseudomonas,* and *Burkholderia genera* were assumed to be responsible for the simultaneous removal of organic carbon and nitrogen. Bioreactor operated at a hydraulic retention time of 12 h, under organic loading charges 0.6 to 16.3 kg TOC m^3 day^{-1}. TOC removal rate varied between 65% and 97% and the total reduction of COD reached 92% without initial pH adjustment. The removal of total Kjeldahl nitrogen for loading charges of 0.5 kg N m^3 day^{-1} reached 75%. However, pH increased during the experiments and caused biofilm separation and a decrease of attached solids concentration, which consequently reduced the carbon and nitrogen removal. When pH was adjusted to 7.5, nitrogen removal improved to 85% at a loading charge of 1 kg N m^3 day^{-1} [83].

A pilot-scale submerged aerobic biofilter (SAB) with a working volume of 178 L and packed with polyethylene corrugated Racshig rings was used in reference [68]. Compressed air was used for aeration. It was continuously operated, with a hydraulic retention time of 24 h, and inoculated with activated sludge biomass. The co-treatment of domestic wastewater with a large content of biodegradable organic matter, and old landfill leachate with high total ammonium nitrogen (TAN) and extremely low BOD/COD ratios, was evaluated. The leachate volumetric ratios 2 and 5 v/v.% were tested. The best results were obtained at a volumetric ratio of 2 v/v.%, where 98% of the BOD, 80% of the COD and DOC, and 90% of the total suspended solids (TSS) were removed. Here, the poorly biodegradable organic matter in leachate was removed by partial degradation. When leachate was added at a volumetric ratio of 5 v/v.%, biodegradation of the low biodegradable organic matter was less efficient and its concentration decreased primarily as a result of dilution. The total ammonium nitrogen was mostly removed (90%) by nitrification.

Generally, the treatment efficiency of the landfill leachate treatment is evaluated according to operational parameters of the investigated treatment plant, and on the basis of analyses of landfill leachate prior and after treatment. Non-specific parameters such as COD, BOD_5, and DOC are routinely determined, as well as the metallic content, concentrations of different ions and some organic pollutants. However, the identification of particular problematic contaminants (pesticides, personal care products, pharmaceuticals, endocrine disrupters, etc.) is difficult and impractical because of analytical limitations (low concentrations, complex extraction methods before analytical procedure, reactivity of the components, etc.), the uncertainty surrounding their bioavailability and the complexity of leachates [67], including possible additive, antagonistic, and synergistic effects of contaminants. As a result, there is a lack of studies dealing with complete determination of reduction of the landfill leachate hazardous environmental impact, which is most suitably determined by the battery of biotests [84]. Only a few studies addressing this problem could be found [56, 67, 70, 85].

4.1.2. Treatment with activated sludge under anaerobic conditions

The anaerobic biodegradation plays an important role in different environmental compartments, such as eutrophic lakes, soils, or sediments, while anaerobic digestion technology for waste and wastewater treatment as well as soil remediation is growing worldwide because of its economic and environmental benefits [12, 46]. The most favorable property of the anaerobic process is the biogas production, contributing to the renewable energy generation. Aerobic systems are suitable for the treatment of low strength wastewaters (BOD_5 < 1,000 mg L^{-1}), while anaerobic systems are suitable for the treatment of highly polluted wastewaters (BOD_5 > 4,000 mg L^{-1}), which is usually not the case when stabilized leachates are discussed. At the same time, after the initial aerobic (acetogenic) phase, landfills actually become anaerobic digesters by themselves [2]. The leachate produced after this phase has already been subjected to anaerobic digestion, so there is little additional treatment efficiency obtained by anaerobic treatment of such leachates. At the same time, anaerobic systems produce an effluent still containing a very high concentration of ammonium nitrogen, which needs a further aerobic stage, necessary to nitrify the anaerobic effluent to be suitable for a watercourse discharge [74, 86]. On the other hand, no such second stage is needed for the aerobic process [11]. Thus, the use of anaerobic-aerobic processes can lead to a reduction in operating costs compared to aerobic treatment alone, while simultaneously resulting in higher organic matter removal efficiency, efficient removal of nitrogen, and a lower waste sludge production. Anaerobic-aerobic systems have received a great deal of attention over the past few decades due to their numerous advantages, not only with regard to the municipal wastewater, but also the sanitary landfill leachates (see Chapter 3.2). Aerobic-anaerobic systems incorporate advantages of both approaches [59]. They could be integrated bioreactors with or without physical separation of aerobic-anaerobic zones, the zones could be switched due to the sequencing mode of operation or they could employ combined culture of anaerobic and aerobic microorganism [74]. They usually achieve more than 70% of COD removal in a short hydraulic retention time (hours-days).

Anaerobic degradation of wastewaters is a very complex and dynamic system, where micro-biological and physico-chemical aspects are strongly linked. This is the reason why granular anaerobic sludge is often applied in various treatment processes, allowing higher loading rates in comparison to conventional systems with dispersed sludge. One of the examples is also the upflow anaerobic sludge blanket reactor (USBR), where wastewater is flowing through a dense bed of sludge with high microbial activity. Granules, which are formed due to the natural self-immobilization of anaerobic bacteria, have a diameter of 1-4 mm. The system could be affected by the presence of suspended and colloidal components of the influent, such as fats, proteins, or cellulose, but these components are usually not typical for landfill leachates. USBR system is well known by its high biomass concentration, high organic loading rates and short hydraulic retention times, a lack of bed clogging, low mass transfer resistance, and large surface area. Another version of the USBR reactor is the expanded granular sludge bed reactor (EGSB) with a high upflow liquid velocity above 4 m s^{-1} and a large height/diameter ratio (> 20) to intensify mixing [74].

Some of the typical treatment systems with prevailing anaerobic conditions, if not completely anaerobic, are presented in Table 6. The systems are in design more or less similar to aerobic ones, presented in Table 5. Anaerobic rotating biological reactors (Table 5, Table 6) are comparable to aerobic ones; they are only covered to avoid contact with air. In both systems there is a series of rotating discs, partly or completely immersed in a reactor through waste-water flows. The system is not energy demanding and it is able to deal with a wide range of flows [74].

Systems with suspended-growth biomass	Systems with attached/immobilized biomass
Activated sludge (AS): · Continuous flow reactors · Sequencing batch reactors (SBR)	Filters: · Upflow system · Downflow system
Membrane bioreactors (MBR): · External membrane module · Submerged/Immersed membrane module	Fluidized bed: · Sand carrier of biomass · Activated sludge carrier of biomass
	Upflow anaerobic sludge blanket (UASB): · Expanded granular sludge bed (EGSB)
	Moving bed bioreactors (MBBR) · Rotating biological contactors

Table 6. Typical treatment system with anaerobic microorganisms in landfill leachate treatment [2, 74, 86].

A possibility of biological treatment in an anaerobic submerged membrane bioreactor was studied in reference [87]. The treatment efficiency under different feeding conditions with different dilution rates of the stabilized leachate and synthetic wastewater (5-75 v/v%) was studied. It contained 2,800-5,000 mg L^{-1}, 1,950-3,650 mg L^{-1}, and 751-840 mg L^{-1} of COD, chloride, and ammonium, respectively. The capacity of the reactor was 29 L and it contained submerged membrane bioreactor with the capillary ultrafiltration module. Reactor was fed with granular sludge, obtained from industrial wastewater treatment plants, and experiments were carried out at 35°C. The effluent form anaerobic reactor was then further treated using reverse osmosis. Treatment was the most efficient at 20 v/v.% of landfill leachate and the system was able to remove up to 90% of COD. For leachate concentration above 30 v/v.%, significant decrease of anaerobic treatment efficiency was observed, probably due to the toxicity of the landfill leachate.

The combination of anaerobic and aerobic reactors was employed in reference [88]. Here, the anaerobic sequencing batch reactor (ASBR) and the pulsed sequencing batch reactor (PSBR), both with 10 L of working volume, were combined to enhance COD and nitrogen removal from the fresh landfill leachate. Anaerobic and aerobic activated sludges from wastewater treatment plants were used to inoculate ASBR and PSBR, respectively. In ASBR, the organics from raw leachate were mainly degraded. During the 157 days long joint operation period, 89.6%-96.7% of COD and 97.0%-98.8% of total nitrogen (TN) removal were achieved. In the effluent, COD and TN were less than 910 mg L^{-1} and 40 mg L^{-1}, respectively, without any extra carbon source addition. Most of the organics in the raw leachate were used as the carbon source during denitrification. In addition, excess organic polymers such as polyhydroxybutyrate (PHB) and glycogen, stored in biomass, acted as the internal carbon source during endogenous denitritation, confirming the possibility of nitrogen removal without the addition of an extra carbon source. These systems are recently more and more often applied, they stop nitrification at the nitrite stage (nitritation), followed directly by reduction to N_2 in anoxic conditions with carbon addition (denitritation). Nitritation/denitritation is attractive because it reduces up to

25% of the total oxygen requirements at the wastewater tretment plant and thus it could significantly reduce costs.

Another reactor system with the up-flow anaerobic sludge bed (UASB) reactor (working volume 3 L) and a 9-L sequencing batch reactor (SBR) in series was used to treat the landfill leachate, in order to enhance the organics and nitrogen removal [89]. The UASB reactor was inoculated with the anaerobic granulated sludge from the methanogenic reactor at wastewater treatment plant, while the aerobic activated sludge from the wastewater treatment plant was used to seed the SBR. Inhibition of the free ammonia on nitrite-oxidizing bacteria and process control were used to achieve the nitrite pathway in the SBR. During a 623 day long experiment, the maximum organic removal rate in the UASB and the maximum ammonium oxidization rate in the SBR were 12.7 kg_{COD} m^{-3} d^{-1} and 0.96 kg_N m^{-3} d^{-1}, respectively. COD, TN, and NH_4^+-N removal efficiencies were 93.5%, 99.5%, and 99.1%, respectively. In the SBR, the nitrite pathway was initiated at low temperatures (14.0°C-18.2°C) and was maintained for 142 days at temperatures 9.0-15°C. Here, stable nitritation was predominantly done by the ammonia-oxidizing bacteria.

An anaerobic pilot-scale sequence batch biofilm reactor (AnSBBR) at room temperature to treat stabilized leachate from a 12-year-old landfill with two extensions, 2 and 5 years, respectively, was used in reference [90]. Leachate was collected from these two extensions. Its COD was 8,566±2,662 mg L^{-1}, with pH around 7.95. The volume of the reactor was 746 L. It was filled with foam cubes (4 x 4 cm with density of 23 g L^{-1}) as inert biomass support. 110 L of the biomass obtained from the existing stabilization pond was used as inoculum. The reaction time was in a range of 5-7 days with filling time of 15 minutes, while 30 minutes were used for the emptying of the system. The treatment efficiency reached over 70% of COD. The authors also studied the kinetics of the process and confirmed that the AnSBR reactor can be considered as a good alternative for the pretreatment of landfill leachate, if it is good or at least partially biodegradable.

It can be concluded that the selection of the biological treatment of the landfill leachate is dependent upon many factors and that the techniques, developed at particular site, could not be always efficiently applied elsewhere [91]. This is also one of the reasons for the intensive development of novel concepts, where fungal treatment seems to be one of viable options.

4.2. Fungal treatment

In the last few decades, the white rot fungi have showed great potential for the removal of hazardous and toxic pollutants. They produce various extracellular ligninolytic enzymes, including laccase (Lac) and manganese peroxidase (MnP), which are involved in the degradation of lignin and their natural lignocellulosic substrates. However, these enzymes are even capable to degrade various pollutants such as phenols, pesticides, polychlorinated biphenyls, chlorinated insecticides, organic dyes, and a range of other compounds. They have been mostly applied for treatment of textile wastewaters due to their excellent decolorization and detoxification effect [14]. A few years ago, a successful treatment of a young landfill leachate with different strains of white rot fungi was presented [14], while the fungal treatment of leachate generated in old landfills has not been investigated so far.

4.2.1. Treatment under fungal growth conditions

In general, two strategies can be applied in water pollutant degradation: (i) direct degradation by active biomass in one reactor or (ii) use of extracted enzymes from the culture medium. In the first case, the fungal growth and the enzyme synthesis, as well as the enzymatic degradation of the pollutant take place in the water solution of the pollutant that is in the wastewater. Here, the effect of the pollutant on the microbial growth and the enzyme synthesis must be taken into account. Fungal enzymes can be intracellular or extracellular products, synthesized during growth or after the growth phase. The pollutant depletion from the wastewater can happen due to the enzymatic degradation or only due to the adsorption on the biomass. Under aerobic conditions, aeration of the reactor is necessary; while under anaerobic conditions, methane is produced. Therefore, the reactor for the first strategy should be considered as a gas-liquid-solid phase system with proper mixing and aeration, since microbial biomass, especially when it is immobilized, can be treated as a solid phase [92].

4.2.2. Treatment with fungal enzymes

From the reactor design point of view, a much simpler configuration, which is only a liquid phase system with proper mixing, can be applied in the second strategy, when only enzyme is to be added to the wastewater. Here, the effect of the pollutant on the enzyme inhibition should be considered [92].

4.2.3. Factors affecting fungal activities

The fungal growth and enzyme production are influenced by numerous factors. *Media composition* has enormous effect on the fungal growth and the production of their degradation systems. In general, a special attention has to be focused on the carbon and nitrogen sources, together with mineral nutrients and other additives. The composition of landfill leachates varies with location and especially with age and is consequently a result of aerobic and later anaerobic conditions in the main body of the landfill.

A *carbon source* is necessary for the growth and enzyme production. In the research during fungal cultivation studies, organic compounds such as glucose, sucrose, starch, and similar have been used. A young landfill leachate usually contains highly biodegradable volatile fatty acids, while in an old landfill leachate, refractory humic and fulvic acids are present, and biodegradable carbon source in various forms must be added to allow fungal growth and enzyme production. A landfill leachate usually also contains toxic phenols and xenobiotics. The white rot fungi can use inorganic as well as organic *nitrogen sources*. Nitrogen demands for growth and especially enzyme production differ markedly between fungal species. In the case of *P.chrysosporium*, production of ligninolytic enzymes is more effective under the conditions of nitrogen limitation, while *B.adusta* produces more LiP and MnP in nitrogen-sufficient media. A landfill leachate usually contains higher concentrations of inhibitory ammonium nitrogen, which must be considered during research. All microorganisms have certain requirements for other medial components, such as mineral nutrients. For example, the white rot fungi need iron, copper, and manganese. However, besides the mentioned metals, a landfill leachate usually contains excess concentrations of toxic heavy metals, which suppress microbial activities [93].

The majority of the filamentous fungi, along with the white rots, grows and produces enzymes optimally at acidic pH values. However, one must distinguish between the optimum pH for growth and enzyme production, the optimum pH for the action of isolated enzymes, and the optimal pH for pollutant degradation [93].

Temperature has to be considered from its influence on the growth and enzyme production, the enzymatic action and the temperature of the waste stream. Most white rot fungi are mesophiles with the optimal cultivation temperature of 27°C-30°C, while optimal temperatures for enzyme reactions are usually higher, but below 65°C [14, 93].

Ligninolytic fungi are obligate aerobes and therefore need oxygen for growth and ligninolytic enzyme synthesis. The oxygen demand depends on the fungus and their ligninolytic system. In addition, leachate treatment also requires oxygen. The problematic of the oxygen supply to the culture media during cultivation has been covered in numerous papers. The major problem is the low oxygen solubility in water (only 8 mg L^{-1} at 20°C). To enhance the oxygen gas-liquid mass transfer and to satisfy the microbial oxygen requirements during cultivation, aeration and agitation are necessary. This can affect the fungal morphology and lead to the decreased rate of enzyme synthesis [94]. As a result of this, various types of bioreactors have been designed, generally divided into static and agitated configurations. The choice of the reactor depends on the particular system, although the appropriate gentle agitation gives as good or even better results as those from static conditions [14, 95].

It is important to optimize an initial leachate concentration for successful degradation in terms of BOD, COD, and ammonium nitrogen removal, as well as detoxification. Namely, leachates are usually toxic to the microorganisms, while the toxicity depends on the type and age of the leachate. This problem can be easily solved by mixing with other types of the wastewaters, mainly the municipal sewage, as already discussed in Chapter 3.2.

4.2.4. Reactors for fungal treatment

A variety of reactor configurations can be used, similar to those for fungal cultivation under submerged conditions. In general, reactors may be divided into two general groups, relying on the situation with the microbial cells. With the first type, the cells are kept in suspension, freely or attached to some support, where they grow and perform their metabolic activities. Suspension can be maintained by a mechanical device, such as a stirrer, through the sparging of air or both. Gentle mixing and aeration have usually been the necessary prerequisites for a successful biomass growth and enzyme production. These reactors are mainly used for the production of biomass or metabolites in various brands of biotechnology, and also applied as activated sludge systems in wastewater treatment technologies. On the other hand, there are many kinds of reactors in the group, where the cells are immobilized. Here, the cells are attached on some type of support, which is in a fixed position within the reactor volume. In these reactors, the attached and active biomass is used for a certain purpose, such as removal of the pollutants from the wastewater. It is shown that batch and continuous operations are effective, both having advantages and disadvantages. Reports in several papers have demonstrated that the repeated use of mycelia over several cycles of decolorization can last from several weeks to a few months [14, 95].

Most of the studies were done under aseptic conditions, while some were effective also during non-aseptic conditions. Toxicity of the leachate strongly affects the degradation and decolorization. For a successful design of a process with the given capacity, the reactor shape and size as well as its operation mode must be selected, and operating conditions, such as concentrations, flow rates, temperature, and pH, must be defined. For the calculation of the necessary pollutant conversion, the degradation reaction rate must be estimated on the basis of literature data or laboratory experiments. This allows for the calculation of time in the batch mode or flow throughput in continuous mode. Typical bioreactors for bioremediation purposes are presented in Table 7 [96, 97].

Bioreactors with suspended-growth biomass	Bioreactors with attached/immobilized biomass
Stirred tank	Packed bed
Bubble column	Trickling filter
Fluidized bed (activated sludge)	Fixed film
	Rotating biological contactor

Table 7. Typical fungal bioreactors for removal of pollutants [96, 97].

4.2.5. *Review of landfill leachate treatment studies with fungi*

Various experiments were done in different types of cultivation vessels. Laboratory equipment, such as multiwell plates and Erlenmeyer flasks, as well as packed bed aerated columns with immobilized culture, was used in the experiments. The aim was mainly to reduce COD, color, and toxicity with various fungal cultures, mostly under sterile conditions. However, the experiments were done with pure and diluted landfill leachates, while the effects of ammonium-N and additional substrate were also investigated. The experiments and their results are presented in Table 8 and briefly discussed in the paragraphs that follow.

Four ecotoxicological assays were used to asses the toxicity of a raw landfill leachate and a mixture of a raw landfill leachate and effluent coming from traditional wastewater treatment plant. All tested organism indicated that both samples exceeded the legal threshold value, showing the ineffectiveness of activated sludge treatment in the reduction of toxicity. To investigate the potential of fungal enzymes for bioremediation, the autochthonous mycoflora of the two samples was evaluated by filtration, because they are already acclimatized to their toxic environment. Ascomycetes showed to be the dominant fraction in both samples, followed by basidiomycetes. However, the samples may represent a risk for human health, since some emerging pathogens were present. A decolorization screening with autochthonous fungi was done with both samples in the presence or absence of glucose. Eleven fungi basidiomycetes and ascomycetes gave promising mycoremediation results by achieving up to 38% decolorization yields [98].

A biological process of landfill leachate treatment with the selected fungal strains on the basis of experiments under sterile conditions in Erlenmeyer flasks was developed. Samples with different concentrations (10%-100%) of young landfill leachates (LFL) with high contents of organic matter, ammonia, salts, heavy metals, phenols and hydrocarbons were treated with *Trametes trogii, Phanerochaete chrysosporium, Lentinus tigrinus* and *Aspergillus niger*. COD removal efficiencies for *P. chrysosporium, T. trogii* and *L. tigrinus* were of 68%, 79% and 90%, respectively, with a two-fold diluted LFL. COD reductions were accompanied by high-enzyme secretion and toxicity reduction, expressed as percent bioluminescence inhibition (< 20%). The effluent was toxic to these strains at LFL concentrations higher than 50% and caused growth inhibition. On the other hand, *A. niger* showed to tolerate raw LFL, since it grew in this media, compared to other tested strains. However, this strain is inefficient in removing phenols and hydrocarbons since toxicity reduction was very low [15].

Landfill leachates, containing high concentrations of phenols and hydrocarbons as well as ammonium nitrogen, and exhibiting high COD and high toxicity, were treated in laboratory experiments by selected strains of white rot fungi (*Trametes trogii* and *Phanerochaete chrysosporium*), with the aim of landfill leachate detoxification. Since high amounts of ammonium nitrogen are problematic for microorganisms, its effect on mycelia growth and enzyme secretion was studied first by adding NH_4Cl to the growth media. Results obtained during the 14 days of cultivation showed that ammonium chloride was not problematic below 2 g L^{-1}, while at 5 g L^{-1}, inhibition of growth and enzyme secretion occurred. Experiments with *T. trogii* and *P. chrysosporium* in 50 v/v.% of the leachate showed 79% and 68% COD removal efficiencies. High enzyme secretion for each strain and a reduction in phenols and hydrocarbons concentrations occurred and consequently, important reduction in the toxicity was achieved [99].

Biotreatability of mature municipal landfill leachate with white rot fungus and its extracellular ligninolytic enzymes was studied in Erlenmeyer flasks. Leachates were obtained from one active and one closed regional municipal landfill, which operated for many years. Both leachates were polluted by organic and inorganic compounds. The white rot fungus *Dichomitus squalens* was able to grow in the mature leachate from the closed landfill. Since it utilized present organic matter as a carbon source, the results showed 60% of DOC and COD removal and decreased toxicity to the bacterium *Aliivibrio fischeri*. In the leachate from the active landfill, the growth of the fungus was inhibited. However, when the leachate was treated with crude enzyme filtrate containing extracellular ligninolytic enzymes, 61% and 44% removal of COD and DOC was reached. In addition, a significant detoxification was proved with the bacterium *Aliivibrio fischeri* and plant *Sinapis alba*. Results showed that the fungal and enzymatic treatment is a promising biological approach for the treatment of mature landfill leachates, so this type of research should continue [70].

Color removal of a landfill leachate in a continuous experiment by immobilized *Trametes versicolor* on polyurethane foam was studied in a 3 L aerated column. Initially, batch experiments in Erlenmeyer flasks under sterile conditions were conducted to find the optimum, pH, optimum co-substrate dose, leachate dilution and contact time for microbial growth, and color removal. The same immobilized fungi under optimal conditions were used after 4 and 15 days of initial growth in several 5-day cycles to test the reuse of fungi. Experiments with the diluted leachate were done to check the effect of organic loading on color removal. Glucose was used as a co-substrate. Results show that that the same immobilized fungi can be reused for at least four cycles, each 5 days long. By using 4-day initially grown biomass, the dilution of leachate did not significantly increase the color removal efficiency without additional glucose. However, about 50% better results in color, BOD and COD removal were obtained by addition of 3 g L^{-1} glucose to the concentrated leachate. Using 15-day initially grown biomass, slightly better color removal was achieved with concentrated leachate. About a twofold increase in color removal in five times diluted leachate was achieved with this biomass, when glucose was added. Results show that *T. versicolor* is capable of treating a highly contaminated landfill leachate and can be considered as a potentially useful microorganism [100].

Trametes versicolor and *Flavodon flavus* immobilized on PUF cubes were used for treatment of landfill leachate. Experiments were done in shaken flasks at room temperature. Effects of pH and co-substrates (glucose, corn starch, and cassava) were investigated at different contact times. Treatment efficiency was evaluated based on color, BOD, and COD removal. For both types of fungi, the optimum pH was 4, the optimum co-substrate concentration was 3 g L^{-1} and the optimum contact time was 10 days. Depending on co-substrate and fungus, the addition of co-substrate at optimum conditions could remove 60-78% of color and reduce 52%-69% of BOD and COD. Promising results prove a potential usefulness of the white rot fungi in the treatment of landfill leachate [101].

Treatment of the landfill leachate by mycelia of *Ganodermaaustrale* immobilized on Ecomat (organic fibers made from oil palm empty fruit brunches), packed in a 0.63 L column, was investigated. Continuous recycling of 1 L of raw leachate and 50 v/v.% diluted leachate at a constant flow (20 mL min^{-1}) was operated for 10 cycles. The results were evaluated on the basis of pH and COD, BOD and ammonium nitrogen removal. Only slight BOD removal was achieved for the raw leachate, while no effect was observed for the diluted sample. COD removal occurred after each cycle with the diluted leachate. Higher COD removal (51%) was observed with the diluted leachate, compared to the raw leachate (23%), after the tenth cycle. Ammonium nitrogen was also reduced after cycle 8 for the diluted (46%) and the raw leachate (31%). The results suggest that the white rot fungus *G. australe* can be considered as a potential candidate in the landfill leachate treatment [102].

Combined fungal and bacterial treatment was also studied. A sequential process in two 10 L laboratory reactors was carried out using a fungal sp. (*Phanerochaete sp.*), followed by a bacterial sp. (*Pseudomonas sp.*) for the degradation and detoxification of contaminants in the landfill leachate. The optimization of cultivation and process parameters for individual fungal and bacterial isolates was done with Box-Behnken design (BBD) and response surface methodology (RSM) for three variables (C source, N source and duration), considering two responses (% of COD and color removal). Treatment in the sequential bioreactor under optimized conditions removed 76.9% of COD and 45.4% of color. In addition, no statistically significant DNA damage of the cells at the end of the treatment was observed, allowing the effluent to be discharged [103].

Type of reactor	Volume	Organism	Type of leachate	Duration	Measured parameters	Removal efficiency	Reference
Multiwell plates	2.5 mL	autochtonous mycoflora	Data not given	20 days	Color	up to 38%	[98]
Erlenmeyer flasks	100 mL	*Trametes trogii, Phanerochaete chrysosporium, Lentinus tigrinus, Aspergillus niger*	Young	15 days	Toxicity COD	up to 80% up to 90%	[15]
Erlenmeyer flasks	100 mL	*Trametes trogii, Phanerochaete chrysosporium,*	Data not given	14 days	COD Toxicity	up to 79% up to 50%	[99]
Erlenmeyer flasks	100 mL	*Dichomitus squalens*	From young landfill From closed landfill	7 days	COD DOC, Toxicity	61% 44% 42%	[70]
Packed bed column, immobilized culture on PFU foam	3 L	*Trametes versicolor*	Young	20 days	Color	up to 78%	[100]
Immobilized culture on PFU foam	100 mL	*Trametes versicolor, Flavodon flavus*	Fresh	15 days	Color BOD COD	up to 78% up to 69% up to 57%	[101]
Immobilized culture	600 mL	*Ganoderma australe*	Data not given	10 cycles (50 min each)	COD NH_3-N	Raw leachate up to 23% Diluted leachate up to 52 % up to 46%	[102]

Table 8. Overview of some fungal landfill leachate treatment studies [15, 70, 98-102].

5. Conclusion and future perspectives

Many municipal as well as industrial wastes still end at the landfill, and consequently the amount of deposited wastes is significant worldwide. Even that waste separation is increasing, organics are still in the landfill and provide good environment for microbial processes, resulting in biogas and leachate production. The composition and the amount of leachate vary with the age and are dependent upon many factors. Significant components of the leachate are organic compounds, which are degradable at the beginning of landfill operation and become more and more persistent and potentially hazardous during the biotic and abiotic processes in the landfill body. These toxic substances should be properly removed to prevent the environmental pollution.

Besides the physico-chemical and chemical methods of leachate treatment, the biological treatment in the form of recycling and combined treatment with domestic sewage, as well as bacterial treatment with the activated sludge under aerobic and anaerobic conditions, have gained significance in the last decade. In addition, the treatment using white rot fungi and their extracellular enzymes seems to be a promising method for the removal of biodegradable and refractory organic matter from the landfill leachate. Before application to the industrial scale (scale-up), research in laboratory and pilot-plant scale are usually required from the engineering point of view. Taking this into account, laboratory toxicity and biodegradability tests are already "a must" at the beginning of the leachate treatment process for both bacterial and fungal processes. However, the activated sludge processes are more than a step forward when compared to fungal processes. From the literature review, experiments with the active sludge have been done in pilot plant reactors with volumes from several liters and a few hundreds of liters, up to lagoons with 60-80 m^3 in non-sterile conditions, which are close to an economically justified situation. It can be expected that the practical application of bacterial landfill leachate treatment will increase in the near future. On the other hand, much less data from the research with fungi is available. Even if this data is promising in regards to the biodegradability tests, it is obtained mainly in the laboratory experiments, mostly in Erlenmeyer flasks under sterile conditions. Therefore, much more time, measured in a decade or so, will be needed for the transfer to a large scale, probably only for a particular and economically tolerable process. A considerable amount of work has been done in the biological treatment of landfill leachate, but there is still a gap regarding mathematical modeling of this process, which has not gained in significance as it has in other fields of biotechnology. Therefore, this engineering tool should be further investigated and applied.

References

[1] Neczaj E, Okoniewska E, Kacprzak M. Treatment of landfill leachate by sequencing batch reactor. Desalination 2005;185:357-362.

[2] Renou S, Givaudan JG, Poulain S, Dirassouyan F, Moulin P. Landfill leachate treatment: Review and opportunity. Journal of Hazardous Materials 2008;150:468-493.

[3] Insel G, Dagdar M, Dogruel S, Dizge N, Cogkor EU, Keskinler B. Biodegradation characteristics and size fractionation of landfill leachate for integrated membrane treatment. Journal of Hazardous Materials 2013;260: 852-832.

[4] Kurniawan TA, Lo WH, Chan GYS. Physico-chemical treatments for removal of recalcitrant contaminants from landfill leachates. Journal of Hazardous Materials 2006;B129:80-100.

[5] Painter HA. Detailed review paper on biodegradability testing, OECD Guidelines for the testing of chemicals, OECD 1995, Paris.

[6] Nyholm N. The European system for standardised legal test for assessing the biodegradability of chemicals. Environmental Toxicology and Chemistry 1991;10(5): 1237-1246.

[7] ISO 15462. Water quality-Selection of tests for biodegradability 2006.

[8] Lyndgaard-Jorgensen P, Nyholm N. Characterisation of the biodegradability of complex wastes. Chemosphere 1988;17(10): 2073-2082.

[9] Danish EPA. Guidance document for risk assessment of industrial wastewater, Miljoprojekt nr. 298, Eds. Pedersen F, Damborg A, Kristensen P; 1995.

[10] Nyholm N. Biodegradability characterization of mixtures of chemical contaminants in wastewater-The utility of biotests. Water Science and Technology 1996;33(6): 165-206.

[11] Henze M, Harremoës P, Jansen J, Arvin E. Wastewater Treatment. Biological and Chemical Processes. Berlin: Springer-Verlag; 1995. 375 p.

[12] Eckenfelder WWJr, Ford DA, Englande AJJr. Industrial Water Quality. 4th ed. New York: WEFPress; 2009. 956 p.

[13] Orhon D, Babuna FG, Karahan O. Industrial Wastewater Treatment by activated Sludge, IWA Publishing 2009; London: p. 3-65.

[14] Singh H. Mycoremediation. Fungal Bioremediation. Hoboken: Wiley Interscience; 2006. p. 1-28.

[15] Ellouze M, Aloui F, Sayadi S. Detoxification of Tunisian Landfill Leachates by Selected Fungi. Journal of Hazardous Materials 2008;150:642-648.

[16] Yedla S. Modified landfill design for sustainable waste management, International Journal of Global Energy Issues 2005;23(1):93-105.

[17] Eriksson O, Carlsson Reich M, Frostell B, Bjorklund A, Assefa G, Sundqvist JO, Granath J, Baky A, Thyselius L. Municipal solid waste management from a systems perspective, Journal of Cleaner Production 2005;13:241-252.

[18] Korica P, Žgajnar Gotvajn A. Models for Evaluating Prevention of Waste on a Macroeconomic Level. Nova Publishing 2015; New York (In Press).

[19] Council Directive 75/442/EEC, 1975. Available from: http://eur-lex.europa.eu/LexUriServ/LexUriServ.do?uri=CELEX:31975L0442:EN:HTML [Accessed 2015-01-21].

[20] Council Directive 2008/98/EC, 2008. Available from: http://eur-lex.europa.eu/LexUriServ/LexUriServ.do?uri=OJ:L:2008:312:0003:0030:en:HTML [Accessed 2015-05-27].

[21] European Community (EC). Communication from the Commission to the European Parliament, the Council, the European Economic and Social Committee and the Committee of the Regions—Roadmap to a Resource Efficient Europe. COM(2011) 571 final 2011. Brussels.

[22] European Community (EC). Proposal for a decision of the European Parliament and of the Council on a General Union Environment Action Program to 2020, 'Living well, within the limits of our planet'. COM(2012) 710 final 2012. Brussels.

[23] Eurostat. Reference Metadata in Euro SDMX Metadata Structure (ESMS). 2012. Available from: http://epp.eurostat. ec.europa.eu/cache/ITY_SDDS/EN/env_wasmun_ esms.htm [Accessed: 2015-01-11]

[24] EU. Council Directive 1999/31/EC of 26 April 1999 on the landfill of waste. OJ L 182 1999. 1-19.

[25] EEA, 2013: European Environment Agency (EEA). Managing municipal solid waste - a review of achievements in 32 European countries 2013. Avilable from: http://www.eea.europa.eu/publications/managing-municipal-solid-waste [Accessed: 2015-02-27].

[26] IPPC, 2013: Integrated Pollution Prevention Directive (IPPC) 2008/1/EC of the European Parliament and of the Council of 15 January 2008 concerning integrated pollution prevention and control 2008. Avilable from: http://europa.eu/legislation_summaries/environment/waste_management/l28045_en.htm [Accessed: 2015-02-27].

[27] EEAb, 2013: European Environment Agency (EEA). Municipal waste management in Slovenia 2013. Avilable from http://www.e:ea.europa.eu/publications/managing-municipal-solid-waste [Accessed: 2015-02-27].

[28] Klimiuk E, Kulikowska D. Organics removal from landfill leachate and activated sludge production on SBR reactors. Waste Management 2006;26:1140-1147.

[29] Monster M, Samuelsson J, Kjeldsen P, Scheutz C. Quantification of methane emissions from 15 Danish landfills using the mobile tracer dispersion method. Waste Management 2015;35:177-186.

[30] Foo KY, Hameed BH. An overview of landfill leachate treatment via activated carbon adsorption process. Journal of Hazardous Materials 2009;171:54-60.

[31] Worell WA, Vesilind PA. Solid Waste Engineerinng. 2nd ed. Stamford: Cengage Learning; 2012. 395 p.

[32] Zhang Q-Q, Tian B-H, Zhang X, Ghulam A, Fnag C-R, He R. Investigation of characteristics of leachate and concentrated leachate in three landfill leachate treatment plants, Waste Management 2013;33:2277-2286.

[33] Matejczyk M, Plaza GA, Nalecz-Jawecki G, Ulfig K,Markowska-Szczupak A. Estimation of the environmental risk posed by landfills using chemical, microbiological and ecotoxicological testing of leachates. Chemosphere 2011;82:1017-1023.

[34] Kalčikova G. The effect of sustainable waste management on municipal landfill leachate composition and treatment. Doctoral thesis 2013. University of Ljubljana, Faculty of Chemistry and Chemical Technology.

[35] Robinson HD, Maris PJ. The treatment of leachates from domestic wastes in landfills. I. Aerobic biological treatment of a medium strength leachate, Water Research 1983;11:1537-1548.

[36] Kjeldsen P, Barlaz MA, Rooker AP, Baun A, Ledin A, Christensen TH. Present and long-term composition of MSW landfill leachate: A Review. Critical Reviews in Environmental Science and Technology 2002;32(4):297-336.

[37] Williams PT. Waste Treatment and Disposal. John Wiley & Sons, Chichester, 2005; p. 155-234.

[38] Cecen F, Aktas O. Aerobic co-tretmet of landfill leachate with domestic wastewater. Environmental Engi*n*eering Science 2004;21:303-312.

[39] Diamadopoulos E, Samaras P, Dabou X, Sakellarpoulos GP. Combined tretment of leachate and domestic sewage in a sequencing bath reactor. Water Science and Technology 1997;36:61-68.

[40] Woldeyohans AK, Worku T, Kloos H. Treatment of leachate by recirculating through dumped solid waste in a sanitary landfill in Addis Ababa, Ethiopia. Ecological Engineering 2014;73:254-259.

[41] Reinhart DR, Al-Yousufi AB. The impact of leachate recirculation on municipal solid waste landfill operating characteristics. Waste Management Resources 1996;14:337-346.

[42] Hermosilla D, Cortijo M, Huang CP. Optimizing the treatment of landfill leachate by conventional Fenton and photo-Fenton processes. Science of the Total Environment 2009;407:3473-3481.

[43] Lopez A, Pagano M, Volpe A, Di Pinto AC. Fenton's pretreatment of mature landfill leachate. Chemosphere 2004;54:1005-1010.

[44] Ricordel C, Djelal H. Treatment of landfill leachate with high proportion of refractory materials by electrocoagulation: System performances and sludge settling characteristics. Journal of Environmental Chemical Engineering 2014;2:1551-1557.

[45] Strotmann U, Pagga U. A growth inhibition test with sewage bacteria-Results of an international ring test 1995. Chemosphere 1996;32(5):921-933.

[46] Žgajnar Gotvajn A, Tratar-Pirc E, Bukovec P, Žnidaršič Plazl P. Evaluation of biotreatability of ionic liquids in aerobic and anaerobic conditions. Water Science and Technology 2014;70(4):698-704.

[47] ISO 9408. Water Quality-Evaluation in an aqueous medium of ultimate aerobic biodegradability of organic compounds-Method by determining the oxygen demand in a closed respirometer 1999.

[48] ISO 7827. Water Quality-Evaluation in an aqueous medium of ultimate aerobic biodegradability of organic compounds-Method by analysis of DOC 1986.

[49] ISO 9888. Water Quality- Evaluation of the elimination and biodegradability of organic compounds in an aqueous medium-Static test (Zahn-Wellens method) 2000.

[50] USEPA. Methods for Aquatic Toxicity Identification Evaluations: Phase I Toxicity Characterisation Procedures. EPA/600/6-91/003; 1991.

[51] USEPA. Methods for Aquatic Toxicity Identification Evaluations: Phase II Toxicity Identification Procedures for Samples Exhibiting Acute and Chronic Toxicity. EPA-600/R-92/080; 1993a.

[52] USEPA. Methods for Aquatic Toxicity Identification Evaluations: Phase III Toxicity Confirmation Procedures for Samples Exhibiting Acute and Chronic Toxicity. EPA-600/R-92/081; 1993b.

[53] USEPA. Toxicity Reduction Evaluation Guidance for Municipal Wastewater Treatment Plants. EPA 833-B-99-002; 1999.

[54] Isidori M, Lavorgna M, Nardelli A, Parrella A. Toxicity identification evaluation of leachates from municipal solid waste landfills: a multispecies approach, Chemosphere 2003;52:85-94.

[55] Cotman M, Žgajnar Gotvajn A. Comparison of different physico-chemical methods for the removal of toxicants from landfill leachate. Journal of Hazardous Materials 2010;178(1/3):298-305.

[56] Žgajnar Gotvajn A, Tišler T, Zagorc-Končan J. Comparison of different treatment strategies for industrial landfill leachate, Journal of Hazardous Material 2009;162:1446-1456.

[57] Cotman M. Razvoj nove strategije za oceno vplivov ekotoksikoloških potencialov odpadne vode na reko Doctoral thesis 2004. University of Ljubljana, Faculty of Chemistry and Chemical Technology (In Slovene).

[58] Muller GT, Giacobbo A, dos Santos Chiaramonte EA, Siqueira Rodrigues MA, Meneguzzi A, Bernandes AM. The effect of sanitary landfill leachate aging on the biologi-

cal treatment and assessment of photoelectrooxidation as a pre-treatment process. Waste Management 2015;36:177-183.

[59] Yao P. Perspectives on technology for landfill leachate treatment. Arabian Journal of Chemisty 2013 (In press).

[60] Mahmud K, Hossain MD, Shams S. Different Treatment Strategies for Highly Polluted Landfill Leachate in Developing Countries. Waste Management 2012;32:2096-2105.

[61] Silva TFCV, Silva MEF, Cunha-Queda AC, Fonseca A, Saraiva I, Sousa MA, Goncalves C, Alpendurada MF, Boaventura RAR, Vilar VJP. Multistage Treatment System for Raw Leachate from Sanitary Landfill Combining Biological Nitrification-denitrification/solar photo-Fenton/biological Processes, at a Scale Close to Industrial-biodegradability Enhancement and Evolution Profile of Trace Pollutants. Water Research 2013;47:6167-6186

[62] Li HS, Zhou SQ, Sun YB, Feng P, Li JD. Advanced treatment of landfill leachate by a new combination process in a full-scale plant. Journal of Hazardous Materials 2009;172:408-415.

[63] Cassano D, Zapata A, Brunetti G, Del Moro G, Di Iaconi C, Oller I, Malato S, Mascolo G. Comparison of several combined/integral biological-AOPs setups for the treatment of municipal landfill leachate: Minimization of operating costs and effluent toxicity. Chemical Engineering Journal 2011;172:250-257.

[64] Singh SK, Townsend TG, Boyer TH. Evaluation of coagulation ($FeCl_3$) and ion exchange (MIEX) for stabilized landfill leachate treatment and high-pressure membrane pretreatment. Separation and Purification Technology 2012;96:98-106.

[65] Chemlal R, Azzouz L, Kernani R, Abdi N, Grib H, Mameri N. Combination of advanced oxidation and biological proceses for the landfill leachate treatment. Ecological Engineering 2014;73:281-289.

[66] Abbod AR, Bao J, Du J, Zheng D, Luo Y. Non-Biodegradable landfill leachate treatment by combined process of agitation, coagulation, SBR and filtration. Waste Management 2014;34:439-447.

[67] Amor C, De Torres-Socias Peres JA, Maldonado M, Oller IS, Malato S, Lucas MS. Mature landfill leachate treatment by coagulation/flocculation combined by Fenton and solar photo-Fenton processes. Journal of Hazardous Materials 2015;286:261-268.

[68] Ferraz FM, Povinelli J, Pozzi E, Vieira EM, Trofino JC. Co-treatment of Landfill Leachate and Domestic Wastewater Using a Submerged Aerobic Biofilter. Journal of Environmental Management 2014;141:9-15.

[69] Roš M. Respirometry of activated sludge. Technomic Publishing Co. Inc. 1993; Lancester: p15-59.

[70] Kalcikova G, Babic J, Pavko A, Zgajnar Gotvajn A. Fungal and Enzymatic Treatment of Mature Municipal Landfill Leachate. Waste Management 2014;34:798-803.

[71] Anastasi A, Prato B, Spina F, Tigini V, Prigione V, Varese C. Decoulorisation and detoxification in the fungal treatment of textile wastewaters from dyeing processes. New Biotechnology 2011;29(1):38-45.

[72] Araki H, Koga K, Inomae K, Kusuta T, Away Y. Intermittent aeration for nitrogen removal in small oxidation ditches. Water Science and Technology 1990;34:131-138.

[73] Ahmed FN, Lan CQ. Treatment of landfill leachate using membrane bioreactors: A review. Desalination 2012;287:41-54.

[74] Chan YJ, Chong MF, Law CL, Hassell DG. A review on anaerobic-aerobic treatment of industrial and municipal wastewater. Chemical Engineering Journal 2009;155:1-18.

[75] Mehmood MK, Adetutu E, Nedwell DB, Ball AS. In situ microbial treatment of landfill leachate using aerated lagoons. Bioresource Technology 2009;100:2741-2744.

[76] Yalcuk A, Ugurlu A. Comparison of horizontal and vertical constructed wetland systems for landfill leachate treatment. Bioresource Technology 2009;100:2521-2526.

[77] Wei Y, Ji M, Li R, Qin F. Organic and nitrogen removal from landfill leachate in aerobic granulat sludge sequening batch reactor. Waste Management 2012;32:448-455.

[78] Trabelsi I, Dhifallah T, Snoussi M, Ben YA, Fumio M, Saidi N, Jedidi N. Cascade Bioreactor with Submerged Biofilm for Aerobic Treatment of Tunisian Landfill Leachate. Bioresource Technology 2011;102:7700-7706.

[79] Canciani R, Emrodi V, Garvaglia M, Malpei F, Passineti E, Buttigklieri G. Effect of oxygen concentration on biological nitrification and microbial kinetics ina cros-flow membrane bioreactor (MBR) and moving-bed biofilm reactor (MBBR) treting old landfill leachate. Journal of Membrane Science 2006;286 202-212.

[80] Galvez A, Giusti L, Zamorano M, Ramos-Ridao AF. Stability and efficiency of biofilms for landfill leachate treatment. Bioresource Technology 2009;100:4895-4898.

[81] Stephenson T, Pollard SJT, Cartell E. Feasibility of biological aerated filters (BAFs) for treating landfill leachate, Conference Proceedings, 9th International Waste Management and Landfill Symposium, S. Margarita di Pula, Cagliary, 2003.

[82] Loukidou MX, Zouboulis AI. Comparison of two biological treatment processes using attached-growth biomas for sanitary landfill leachate treatment. Environmental Pollution 2001;111:273-281.

[83] Trabelsi I, Sellami I, Dhifallah T, Medhioub K, Bousselmi L, Ghrabi A. Coupling of anoxic and aerobic biological treatment of landfill leachate. Desalination 2009;246:506-513.

[84] Kalčikova G, Vavrova M., Zagorc-Končan J, Žgajnar Gotvajn A. Evaluation of the hazardous impact of landfill leachates by toxicity and biodegradability tests. Environmental Technology 2011;32(2): 1345-1353.

[85] Billa DM, Montalvao AF, Silva AC, Dezoti M. Ozonation of landfill leachate: Evaluation of toxicity removal and biodegradability improvement. Journal of Hazardous Materials 2005;11(2/3):235-242.

[86] NTUA National Technical University of Athens. Bioreactors for metal bearing wastewater treatment. Available from: http://www.metal.ntua.gr/~pkousi/e-learning/bioreactors/page_18.htm. [Accessed: 2015-01-28].

[87] Bohdziewicz J, Neczaj E, Kwarciak A. Landfill leachate treatment by means of anaerobic membrane bioreactor. Desalination 2008;221:559-565.

[88] Zhu R, Wanga S, Li J, Wang K, Miao L, Ma B, Yongzhen Y. Biological Nitrogen Removal from Landfill Leachate Using Anaerobic-aerobic Process: Denitritation via Organics in Raw Leachate and Intracellular Storage Polymers of Microorganisms. Bioresource Technology 2013;128:401-408.

[89] Sun H, Shi, Peng Y. Advanced Treatment of Landfill Leachate Using Anaerobic-aerobic Process: Organic Removal by Simultaneous Denitritation and Methanogenesis and Nitrogen Removal via Nitrite. Bioresource Technology 2015;177:337-435.

[90] Contrera RC, Da Cruz Di Silva K, Morita MD, Domingues Rodrigues JA, Zaiat M, Schalch V. First-order kinetics of landfill leachate treatment in a pilot-scale anaerobic sequence by batch biofilm reactors. Journal of Environmental Management 2014;145:385-393.

[91] Cortez S, Teixeira P, Oliveira R, Mota M. Landfill leachate Treatment: A Portugese case study, In: Cabral GBC, Botelho BAE, editors. Landfills Waste Management, Regional Practices and Environmental Impact. New York: Nova Science Publishers Inc; 2012; p. 83-136.

[92] Pavko A. Fungal Decolourization and Degradation of Synthetic Dyes. Some Chemical Engineering Aspects. In: Garcia Einschlag FS, editors Wastewater Treatment and Reutilization. Rijeka: InTech; 2011. p. 65-88.

[93] Knapp JS, Vantoch-Wood EJ, Zhang F. Use of Wood-rotting fungi for the decolorization of dyes and industrial effluents. In: Gadd GM. (Ed.) Fungi in Bioremediation. Cambridge: Cambridge University Press; 2001. p. 253-261.

[94] Žnidaršič P, Pavko A. (2001). The morphology of filamentous fungi in submerged cultivations as a bioprocess parameter. Food Technology and Biotechnology 2001;39:237-252.

[95] Doran PM. Bioprocess Engineering Principles. San Diego: Academic Press; 1995. p. 333-391.

[96] Martin A. Biodegradation and Bioremediation. London: Academic Press; 1999. p. 355-376.

[97] Moreira MT, Feijoo G, Lema JM. Fungal Bioreactors: Applications to White-rot Fungi. Reviews in Environmental Science and Biotechnology 2003;2:247-259.

[98] Tigini V, Prigione V, Varese GC. Mycological and Ecotoxicological Characterisation of Landfill Leachate before and after Traditional Treatments. Science of the Total Environment 2014;487:335-341.

[99] Ellouze M, Aloui F, Sayadi S. Effect of high Ammonia Concentrations on Fungal Treatment of Tunisian Landfill Leachates. Desalination 2009;246:468-477.

[100] Saetang J, Babel S. Effect of Leachate Loading Rate and Incubation Period on the Treatment Efficiency by *T. versicolor* Immobilized on Foam Cubes. Int. J. Environ. Sci. Tech. 2009;6(3):457-466.

[101] Saetang J, Babel S. Fungi Immobilization for Landfill Leachate Treatment. Water Science and Technology 2010;62.6:1240-1247.

[102] Noorlidah A, Wan Razarinah WAR, Noor ZM, Rosna MT. Treatment of Landfill Leachate Using *Ganoderma Australe* Mycelia Immobilized on Ecomat. International Journal of Environmental Science and Development 2013;4(5):483-487.

[103] Ghosh P, Thakur IS. Enhanced removal of COD and color from landfill leachate in a sequential bioreactor. Bioresource Technology 2014;170:10-19.

Chapter 6

Wet Air Oxidation of Aqueous Wastes

Abstract

Wet air oxidation (WAO) is a key technology in the disposal of industrial and agricultural process wastewaters. It is often used coupled with activated sludge treatment at a wastewater treatment plant (WWTP) as preliminary conversion of toxic and/or non-biodegradable components. The process is based on a high temperature and pressure reaction of the oxidizable materials in water with air or oxygen, in most cases in a bubble column reactor. The oxidation is a chain type radical reaction. The intensification of this technology is possible with the application of homogeneous and heterogeneous catalysts, recently non-thermal radical generating methods (UV/H_2O_2, ozonization, Fenton type processes) gathered ground also. The most frequent use of the process is in sludge treatment and oxidation of spent caustic of refineries or ethylene plants.

Keywords: Process wastewater treatment, wet air oxidation, catalysts, intensification of wet oxidation

1. Introduction

Nowadays, many industries and some agricultural activities generate large quantities of aqueous wastes containing organic and inorganic substances known as process wastewaters (PWWs). Before discharging these wastewaters into the environment like domestic wastewaters, industrial wastes must undergo various physical, biological, chemical, or combined treatments to reduce their toxicity and to convert them into biodegradable materials.

Liquid wastes of high organic and inorganic content have been classified as hazardous wastes and are mainly disposed of by incineration. Incineration is typically used when everything else fails and then the hazardous and polluting components are eliminated with fuel consumption, which bears high costs while producing secondary pollutants such as dioxin, furans, and sulfur dioxide.

Among typical treatment methods applied for these kinds of wastes, biological treatment is the primary method for removal of organic pollutants, but often not suitable since they may contain toxic, non-biodegradable, and hazardous pollutants. Moreover, microbes are vulnerable to (chemical) shocks, further limiting their use in the chemical industry. Some wastewater streams are too concentrated to be cleaned effectively by biological treatment. On the other hand non-biodegradable liquid wastes generated from industrial and agricultural processes can be treated by chemical oxidation, wet oxidation, or with advanced oxidation processes (AOPs) in order to eliminate their toxicity and enhance their biodegradability.

Wet oxidation, also known as wet air oxidation, refers to a process of oxidizing suspended or dissolved material in liquid phase with dissolved oxygen at elevated temperature. It is a method for treatment of waste streams that are too dilute to incinerate and too concentrated for biological treatment.

2. Thermal wet oxidation processes

2.1. Wet air oxidation

Thermal wet oxidation processes use high-temperature and high-pressure air or oxygen as oxidant. Wet air oxidation (WAO) refers to a process of oxidizing wastewaters, the water containing liquid wastes under pressure with air or oxygen and at high temperature (>120°C). It is a good option for treating the high-organic content PWWs which can originate from fine chemical and pharmaceutical industries. The chemistry of wet air oxidation involves chain reactions of radicals formed from organic and inorganic compounds present in the reaction mixture. In this hydrothermal process, the organic pollutants are converted into easily biodegradable substances or completely mineralized and the inorganic compounds are converted into their form with higher oxidation value, such as sulfides into sulfates.

A typical condition for wet air oxidation ranges from 180°C and pressure of 2 MPa to 315°C and 20 MPa pressure.

2.1.1. Industrial application of wet air oxidation

WAO technologies have been commercialized around 60 years ago. Applying different temperatures, one can mention three different oxidation categories, namely: low temperature oxidation, medium temperature oxidation, and high temperature oxidation.

Commercial application of low-temperature oxidation (100°C-200°C) involves low-pressure thermal conditioning (LPO) of activated sludge and the treatment of low-strength sulfidic spent caustic. The most prevalent WAO applications are for ethylene plant spent caustic and refinery spent caustic.

The ethylene plant spent caustic is traditionally oxidized at low temperatures in the range of 120°C-220°C; the main purpose is to destroy the odorous sulfide content of this effluent. The refinery spent caustic WAO application operates at 240°C-260°C that already fits the medium temperature category, as well as the WAO of some organic wastes. Other industrial wastes treated by low-temperature oxidation include cyanide and phosphorous wastes, as well as non-chlorinated pesticides. High-temperature oxidation 260°C-320°C is used for refinery spent caustic, sludge destruction, and most WAO treated industrial wastewaters. Most organic industrial wastes are oxidized in this temperature range, including pharmaceutical wastes as well as pesticides, solvents, and the complete destruction of liquid wastes of pulp and paper production and other organic sludges.

2.1.2. Zimpro® wet air oxidation and related processes

The history of WAO started almost about 60 years ago, when Zimmermann observed that he could dispose pulp mill liquors using air at high pressure leading to the oxidation of organic compounds dissolved or suspended in water at relatively low temperatures in the presence of oxygen [1].

They took spent pulping liquor from a local paper mill to produce artificial vanilla flavoring (vanillin) by partial oxidation of ligno-sulfonic acids. They perfected the wet air oxidation process (or the "Zimmermann Process" as it was known), and expanded it to other applications, including wastewater treatment. The company, developed and installed these wet oxidation plants, had a diversified history as it was established by Sterling Drug Inc. as "Zimpro" at Rothschild in 1961, building the engineering and research center along the Wisconsin River. After a long expansion period, with new developments and acquisitions Zimpro was purchased by USFilter in 1996. In April 1999, Paris-based Vivendi announced the acquisition of USFilter. Siemens bought USFilter in May 2004. Presently, it is owned by Siemens Co. under the name "Siemens Water Technologies."

Since the beginning, this process (Zimpro) had been mainly used for sewage sludge treatment, but by the early 1970s, it was applied to regenerate spent powdered active carbon from wastewater treatment processes. During 1980s, WAO began to be more useful as an industrial waste treatment technology. Zimpro Products installed the first wet oxidation unit in 1982 to treat ethylene plant spent caustic. The next year, they installed and operated a wet oxidation system in California for the treatment and detoxification of hazardous wastes. In 1992, Zimpro installed a wet oxidation system at Sterling Organics in Dudley, Northumberland, UK, for pharmaceutical wastewater treatment. Currently, about 200 full-scale WAO plants are in operation for the treatment of a variety of effluent streams (municipal sludge, night soil, carbon regeneration, acrylonitrile process effluent, metallurgical coking, ethylene production spent caustic, paper filler, industrial activated sludge, pulping liquor, warfare chemicals, paper mill sludge, explosives, monosodium glutamate production, polysulfide rubber, textile sludge, chrome tannery waste, petroleum refining spent caustic, miscellaneous industrial sludges, nuclear reprocessing wastes) [2, 3]. In 2009, 2011, and 2012, Siemens contracted with Sinopec to build several WAO units in PR China for the disposal of spent caustic from ethylene plants and refineries.

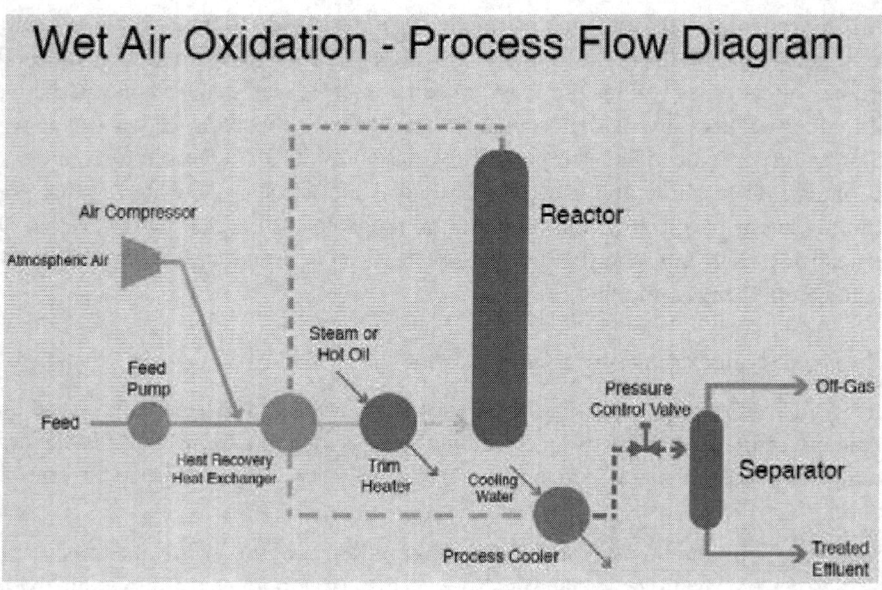

Figure 1. Typical flow diagram of WAO [2]

The typical wet oxidation system (Figure 1) is a continuous process using rotary compressor and pump to compress the air (or oxygen) and feed liquid stream to the required operating pressure. Heat exchangers serve to recover energy from the reactor effluent and use it to preheat the feed/air mixture entering the reactor. Auxiliary energy, usually steam, is necessary for startup and can provide trim heat if required. The residence time in the reactor vessel is several hours at a temperature that enables the oxidation reactions to proceed in some cases toward total mineralization. The reactor is a bubble column; it is coupled after the heat exchanger with a separator for the separation of the effluent and the off-gases. Since the oxidation reactions are exothermic, sufficient energy may be released in the reactor to allow the wet oxidation system to operate without any additional heat input at or above COD > ~10000 mg/L.

The typical reactions during WAO:

$$\text{Organics} + O_2 \rightarrow CO_2 + H_2O + RCOOH^*$$
$$\text{Sulfur Species} + O_2 \rightarrow SO_4^{2-}$$
$$\text{Organic Cl} + O_2 \rightarrow Cl^- + CO_2 + RCOOH^*$$
$$\text{Organic N} + O_2 \rightarrow NH_3 + CO_2 + RCOOH^*$$
$$\text{Phosphorus} + O_2 \rightarrow PO_4^{3-}$$

The oxidation reactions occur at temperatures of 150°C to 320°C and at pressures from 10 bar to 220 bar. The required operating temperature is determined by the treatment objectives. Higher temperatures require higher pressure to maintain a liquid phase in the system. Typical industrial applications for the WAO process have a feed flow rate of 1 m^3/h to 50 m^3/h per unit, with a chemical oxygen demand (COD) from 10,000 mg/L to 150,000 mg/L (higher COD with dilution).

During the decades of process development, huge amount of data were collected and published about the oxidation properties of individual compounds, different process wastewaters, and sludges. In WAO of ethylene spent caustic, the conversion of sulfide is nearly 100% at 200°C during 60 minutes. The high conversion of the polluting compounds in refinery spent caustic needs higher temperature of oxidation (260°C), residuals from pesticide and herbicide production need even higher (280°C). This is associated obviously with higher total pressure because of the increased vapor pressure of water and other volatiles.

A good instance is, for illustrating the technical details and problems of WAO, the oxidation of ethylene plant spent caustic, which is one of best-elaborated technology among WAO processes. Such spent caustic liquor contains as major components the compounds listed in Table 1. The sulfur containing compounds are oxidized to sulfate, being present in the basic solution as sodium sulfate, the organic components are oxidized primarily to carboxylic acids, such as acetic, oxalic, formic, and propionic acid. At 200°C and 28 bar pressure only the partial oxidation of organic compounds occurs, as the forming carboxylic acids are well biodegradable [4].

Compound	**Concentration range**
NaHS	0.5-6 %
Na_2CO_3	1-5 %
NaOH	1-4 %
NaSR	0-0.2 %
Soluble oil	50-150 ppm
TOC	50-1500 ppm
Benzene	20-100 ppm

Table 1. Composition of ethylene plant spent caustic

WAO reliability can be hampered by off-spec feed, which is affected by the upstream processing and handling of the spent caustic. The feed of the WAO plant for treating ethylene plant spent caustic contains some reactive organic compounds (called "red oil," main components are primarily aldehydes that form high molecular weight materials through the so-called aldol condensation reaction) that cause fouling not only in the ethylene compressor and the separation equipments but in the heat exchanger in front of the oxidation reactor serving for the preheating of the caustic feed and cooling of the effluent. The fouling is even worse in the presence of iron that usually forms insoluble scale, which has to be removed by chemical and/ or mechanical treatment. Chloride is dangerous because of corrosion, in spite of the use of special alloys in the WAO equipments.

A special process is the so-called VerTech oxidation of sludges. This applies an underground installation, consisting of concentric tubes as heat exchangers going down to 1200 m depth to an oxidation vessel. Due to the depth of the vessel, the bottom pressure in the reactor is above 100 bar at 275°C without the need of high-pressure pumps at the surface. The building of the installation required drilling and casing technology developed by the gas and oil production industry. The operation of the system required a frequent descaling operation with nitric acid in order to preserve the efficiency of heat exchange in the concentric tubes. The output of this plant was 80 tons of sludge per day [5].

2.1.3. *Kinetic mechanism of WAO*

WAO of organic pollutants is generally described by a free-radical chain reaction mechanism in which the induction period to generate a minimum radical concentration is of great significance [6-9]. In the most detailed studied reaction in WAO, in the oxidation of phenol water solution during the induction period, practically no change was observed in phenol concentration. Once the critical concentration of free radical is reached, fast reaction takes place (propagation step) when almost all phenol is oxidized. It has been found that the induction period length depends on oxygen concentration, temperature, type of organic compound and if applied on the catalyst concentration [2, 10-13]. The pH also has influence on the induction period that actually is shorter for pH values of about 4, and is increasing with the increase in pH [14].

In wet oxidation, the reaction chains are thermally initiated [15]:

$$RH \rightarrow R^{\cdot} + H^{\cdot} \tag{1}$$

$$RH + O_2 \rightarrow R^{\cdot} + HO_2^{\cdot} \tag{2}$$

$$R^{\cdot} + O_2 \rightarrow ROO^{\cdot} \tag{3}$$

$$ROO^{\cdot} + RH \rightarrow ROOH + R^{\cdot} \tag{4}$$

$$RH + (OH^{\cdot}, HO_2^{\cdot}, ROO^{\cdot}) \rightarrow R^{\cdot} + \left(H_2O, H_2O_2, ROOH\right) \tag{5}$$

$$H_2O_2 \rightarrow 2\,HO^{\cdot} \tag{6}$$

In the mechanism, R is any organic molecule present in the reaction mixture. In addition to the organic radicals several inorganic radicals participate in the degradation such as H, HO, and HO_2/O_2^-. Among these radicals HO is the most reactive, it reacts with aromatic molecules (Ph=phenol) in radical addition reaction with practically diffusion limited rate coefficient at all temperatures up to the supercritical value [16].

$$Ph + HO^{\cdot} \rightarrow PhHO^{\cdot} \quad (7)$$

The hydroxycyclohexadienyl radical formed in Reaction (7) from phenol is also highly sensitive to oxygen.

In all real PWW and in the reaction mixtures of wet oxidation in stainless steel autoclaves, Fe ions are present with measurable concentration [17-19]. These ions and other transition metal ions present accelerate the decomposition of peroxide to HO radicals in Fenton-type processes. The steady state (propagation step) is then followed by the third step (termination step) characterized by a slow oxidation rate.

The first step is the chain initiation, in which free radicals (R, OH, HO_2) are produced by dissociation (1) and the bimolecular reaction of dissolved oxygen with the organic compound (2), which is found to be very slow at low temperatures. When the free radical R is formed, it can readily react with molecular oxygen to give peroxoradical (ROO) (3). The other reaction is the formation of hydrogen peroxide, which decomposes with metal catalysis to OH radicals (6). Finally the peroxo radical gives with the parent compound a free radical and hydroperoxide (4). The OH radicals oxidize the parent compound into a free radical again (5).

In the mechanism, the organic parent compound (RH) can react thus with molecular oxygen, the organic peroxyl (ROO) (5), hydroxyl (OH) (6) and hydroperoxyl (HO_2) (3) radicals [20, 21].

Intermediates formation is of great importance in WAO and has been reviewed by Devlin and Harris for the oxidation of aqueous phenol with dissolved oxygen [22]. The conclusion was that, at elevated temperatures, oxygen is capable of three different oxidation reactions with the organic: (i) introducing an oxygen atom into an aromatic ring to form a dihydric phenol or quinone; (ii) attacking carbon to carbon double bonds to form carbonyl compounds; and (iii) oxidizing alcohols and carbonyl groups to form carboxylic acids. The ring compound intermediates (dihydric phenols and quinones) were formed under conditions near the stoichiometric ratio of phenol and oxygen, increasing in quantity when oxygen was in deficiency. The unsaturated acids, namely maleic and acrylic and saturated ones, namely formic, acetic and oxalic appear independently of phenol to oxygen ratio used. Malonic, propionic and succinic acids were identified only in case of deficit of oxygen. Malonic acid undergoes decarboxylation to produce acetic acid and carbon dioxide.

Another interesting research was done about the kinetics of oxidation of phenol, and nine substituted phenols were investigated [10]. The process was studied in a 1 L stainless steel autoclave at temperatures in the range of 150°C-180°C and the initial phenol concentration was 200 mg/L. The oxidation reaction found to be the first order for oxygen and also first order with respect to phenolic substrates in both cases. The overall oxidation reaction rate was found to be kinetically controlled when the temperature was less than 195°C and the phenol concentration was less than 200 mg/L. At higher temperatures (>240°C) and higher phenol concentration (>20000 mg/L) the overall oxidation reaction rate became mass transfer controlled.

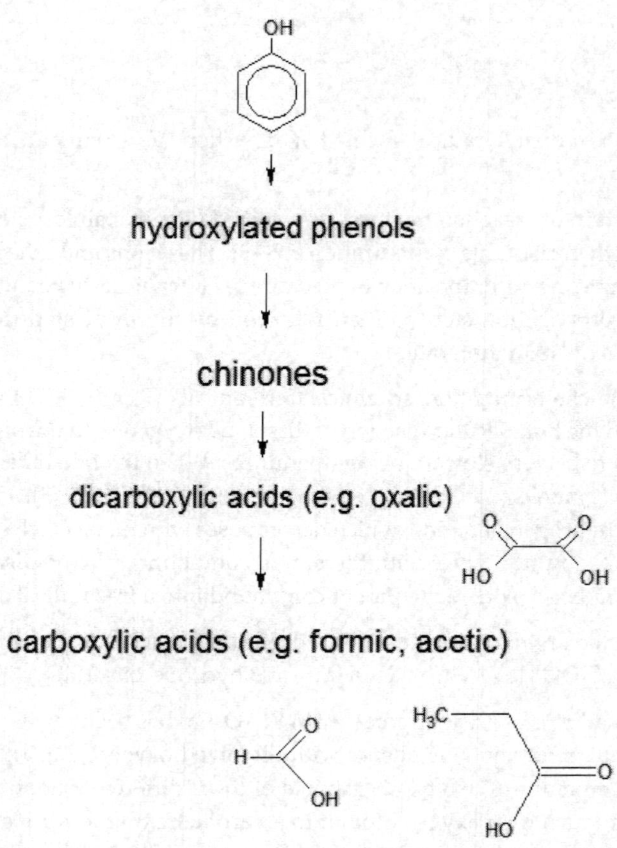

Figure 2. Proposed reaction pathway for phenol oxidation by molecular oxygen.

2.2. Catalytic Wet Air Oxidation (CWAO)

In the 80s significant need was revealed toward treatment of highly concentrated wastewaters of chemical and pharmaceutical production, as well as residual sludge [22, 23]. Aside from WAO, CW(A)O has been applied to many different model effluents, but relatively few works have been devoted to real and complex industrial wastes [2, 24-28].

In order to carry out wet oxidation under milder conditions (at lower temperature and pressure) an alternative way would be catalytic wet (air) oxidation (CW[A]O). Soluble transition metal salts (such as copper and iron salts) have been found to give significant enhancement of the reaction rate, but post-treatment is needed to separate and recycle them.

Heterogeneous catalysts have the advantage that they can be used without the problem of separation and for continuous operation. Mixtures of metal oxides of Cu, Zn, Co, Mn, and Bi are reported to exhibit good activity, but leaching of these catalysts was detected [25, 29-32]

The reaction mechanism of CWAO is thought to be similar to the mechanism of WAO, and the function of the catalyst is essential promoting the formation of free radicals.

Kinetic study was carried out on the phenol oxidation by CWAO using aqueous copper nitrate as homogeneous catalyst. A kinetic model has been established based on the free radical mechanism where the electron transfer from copper to phenol was assumed to initiate the formation of free radicals and this led to propose that the formation of free radicals (PhO· and PhOOO·) is primarily due to the electron transfer from metal to phenol. In this model, the reaction orders were found to be approximately 1.0, 0.5, and 0.5 with respect to phenol, oxygen, and copper concentrations, respectively. In order to verify the proposed kinetics, a series of CWAO experimental tests were done at 313-333 K, oxygen partial pressures of 0.6-1.9 MPa, and copper concentrations 0-13 mg L^{-1}. The experimental data fitted well with the model [33].

With supported copper oxide catalyst in the temperature range of 120°C-160°C, with 0.6 MPa and 1.2 MPa oxygen pressures, it was found that the reaction was first order with respect to phenol concentration and half order with respect to oxygen partial pressure [34].

The composition of the most active heterogeneous catalysts in wet oxidation, namely that they are multicomponent, alludes on the validity of the Mars-van Krevelen mechanism well known in catalytic oxidation chemistry [35, 36].

2.2.1. Industrial homogenous CWAO processes

In these processes, homogeneous transition metal catalysts are used that need, however, to be separated and then recycled to the reactor or discarded.

The first process to be mentioned was developed by ***Ciba-Geigy***, which uses a copper salt as a catalyst. From the oxidized material the catalyst has to be separated as copper sulfide and recycled into the reactor, which is titanium lined. The unit works at 300°C and pressure above 100 bar. Three units that are installed in Germany and Switzerland have achieved high oxidation efficiencies (95%-99%) on chemical and pharmaceutical wastes at elevated temperatures.

The other one is the ***LOPROX*** process, a relatively low-temperature and low-pressure wet oxidation that was developed by Bayer AG for treatment of organic substances, especially aromatic compounds, which degrade too slow in normal biological plants or adversely affect the degradation of other substances. The disposal of aromatics is important for two reasons, they are forming significant portion of PWWs from the chemical industry. They are present in the clarified activated sludge also, which contains humic acids, these have aromatic part with chlorine substituents. It takes place in the presence of oxygen in acidic range in a multi-stage bubble column reactor under relatively mild operating conditions (temperature below 200°C, pressure 0.5-2.0 MPa) the catalyst is the combination of Fe^{2+} ions and quinone-generating substances. The residence time is 1-3 hours. Above COD value of 6-8 g/L the process is autothermal, no heat energy input is needed. A critical issue is the choice of structural materials of the reactor and of other hot parts of the system. Enameled or PTFE lined steel can be used up to 160°C, titanium and titanium-palladium alloys are applied up to 200°C because of the acidic pH. Several LOPROX plants are in operation at Bayer AG. They dispose PWWs from intermediate, dyestuff, pharmaceutical, paper and pulp production, and clarified sludge.

Veolia developed the **ATHOS®** process for the treatment of clarified sludge at WWTPs. It works with Cu ion catalysis at 250°C and 50-60 bar pressure with pure oxygen. The reactor is perfectly mixed, because it has a circulation loop, not the usual bubble column. The heat exchangers are working with extremely high-temperature water as heating source. Such plant is working at the Brussels WWTP, in a complex line for the abatement of clarified sludge [37].

2.2.2. Industrial heterogeneous CWAO processes

The development of stable and active heterogeneous catalysts for WO of PWWs is a difficult task, as the substrates to be oxidized are diverse, the wastes are multicomponent, and severe conditions are needed for the completion of the reactions. At high-temperature and high-oxygen partial pressure even at basic pH of the reaction mixture the leaching of the active component(s) of the catalyst into the water solution frequently occurs. As mentioned in a review, there are two catalytic WAO technologies that have been developed in the late '80s in Japan. Both processes use heterogeneous catalysts, precious metals deposited on titania-zirconia carriers. They are able to oxidize two refractory compounds namely acetic acid and ammonia also [27].

The **NS-LC** process uses a Pt-Pd/TiO_2-ZrO_2 honeycomb catalyst. Typical operating conditions are 220°C and 4 MPa pressure with space velocity = 2 hour^{-1}, which with these operating conditions the oxidation of compounds such as phenol, formaldehyde, acetic acid, glucose, etc., exceeds 99% conversion. In the absence of a catalyst the removal efficiency would go down to 5%-50% [38]. The specialty of the reactor is the segmented gas-liquid flow, which means that the liquid plugs are sandwiched between two gas plugs, and the flow has a mass transfer increasing and solid deposition preventing effect.

The other process, which is called **Osaka Gas,** is based on a mixture of precious and base metals on titania or titania-zirconia carriers in a form of honeycomb or sphere. This process has been applied in several industrial and urban wastes. A typical pilot plant at British Gas's London Research station works at 250°C and pressure of 9 MPa, with 200 L/h feed of waste [39].

Kurita Company developed a process to abate ammonia with the oxidation agent nitrite in the presence of a supported Pt catalyst. The reaction temperature (170°C) is lower than in usual WO.

One of the recent developments is the ***CALIPHOX*** process made by the National Institute of Chemistry of Slovenia and an engineering firm for treatment of industrial wastewaters with a metal oxide catalyst in the extruded form in a trickle-bed reactor. It is operating in relatively mild conditions (180°C, 4 MPa). The catalyst is based on the work of Pintar and Levec [40] who studied CuO- ZnO-Al_2O_3.

2.2.3. *Types of CWAO catalysts*

The objective of catalyst application, beside reaction rate enhancement, is to operate among milder conditions. The catalyst is usually a metal salt, a metal oxide, or the metal itself [Table 2.]. As we know, heterogeneous catalysts based on precious metals deposited on stable supports are less sensitive to leaching [25, 31, 41-44]. Different catalysts were applied and their effects were investigated on different catalytic wet oxidation processes in the past years. Pt and Ru on ceria and zirkonia-ceria supports were tested in the oxidation of acetic acid that was accompanied by the loss of activity [32]. In a following paper, the same authors described the reason for deactivation of the Pt catalyst, the accumulation of carbonate species on the surface [45]. Recently, activity of Ru-oxide on different oxide supports in acetic acid oxidation was reported. The mixed Zr, Ce oxide supported catalyst proved to be the most active [46].

Substrate	*Reaction condition*	*Oxidant, catalyst*	*Reactor type*	*Removal*	*Reference*
Phenol	120-160°C 6-12bar(O_2)	Air, Cu	trickle bed reactor	<90%COD	Fortuny et al.1999
Phenol, acetic acid	170-200°C 20 bar (O_2)	O_2, Ru, Pt, Rh	batch	<97%COD	Duprez et al. 1996
Phenol chlorophenol nitro phenol	150-210°C 30 bar (total)	O_2, CuO, Zn, Co oxides	trickle bed reactor	>95%TOC	Pintar&Levec1994
Phenol	90-150°C	CuO+Al_2O_3 O_2, CuO+ZnO	batch	100% X	Akyurtlu et al.1998
p-chlorophenol	180°C 26bar (total)	O_2,Pt, Pd,Ru	slurry	<98% TOC	Qin et al., 2001
Ammonia	110°C-130°C	O_2, Pt/SDB	trickle bed reactor	100% X	Huang et al., 2001
Ethyl benzene	310-390°C 0-1 bar (O_2)	O_2, AC	packed bed reactor	50%	Pereira, et al, 2000
Aniline	160-230°C 20 bar O_2	O_2, Ru/CeO_2	batch	100% X	Oliviero et al., 2003
Paper industry wastewater	140-190°C 5 bar (O_2)	O_2, Cu/Mn, Cu/Pd, Mn/Pd	batch	>84% TOC	Akolekar et al., 2002
Carboxylic acid	180°C 1-11bar(total)	Air, Pt/Al_2O_3	batch	100% X	Lee & Kim, 2000
Kraft bleach plant effluent	190°C 8 bar (O_2)	Air, Ru/TiO_2	trickle bed reactor	< 90%TOC	Pintar et al., 2001

Table 2. CWAO of organic pollutants and industrial effluents. [40-41, 47-56]

2.2.3.1. *Supported noble metal catalyst*

Various noble metals (Ru, Pt, Rh, Ir, and Pd) and some metal oxides (Cu, Mn, Co, Cr, V, Ti, Bi, and Zn) have traditionally been used as heterogeneous catalysts in CWAO. Several studies have ranked catalysts according to their activity. Imamura and his colleagues ranked noble metal and metal oxide catalysts according to their total organic carbon conversion achieved in 1 h, during the oxidation of polyethylene glycol at 200°C and pH of 5.4 [57]. They found the following order: Ru = Rh = Pt > Ir > Pd > MnO.

2.2.3.2. *Supported metal oxides*

Metal oxides can be classified according to their physical-chemical properties. One of these properties is the stability of metal oxide. Metals with unstable high oxidation state oxides, such as Pt, Pd, Ru, Au, and Ag do not perform stable bulk oxides at moderate temperatures. Most of the commonly used metal oxide catalysts (Ti, V, Cr, Mn, Zn, and Al) have stable high oxidation state oxides. Fe, Co, Ni, and Pb belong to the group with intermediate stability of high oxidation state oxides [58].

Mixtures of metal oxides frequently exhibit greater activity than the single oxide. Cobalt, copper, or nickel oxide in combination with the following oxides of iron (III), platinum, palladium, or ruthenium, are reported as effective oxidation catalysts above 100°C [59]. In addition, combining two or more metal catalysts may improve non-selective catalytic activity.

Metal oxides are usually applied in the form of powders and fine particles, and with this form of catalyst structure we can achieve maximum specific surface area, but the dispersion of the particles can create unsteady state. To keep the stable state of the catalyst, at the same time not losing the active phase, some porous supports can be used. Commonly, alumina and zeolites are used as support, but surface area of aluminum oxide is limited and the pore size of zeolites cannot be permeable for large size organic molecules.

2.2.3.3. *Activated carbon catalysts for CWAO*

Another promising catalyst could be activated carbon (AC) that shows good properties as adsorbent for both organic materials and oxygen because of its porous structure and high surface area [60, 61]. Activated carbon is stable in highly acidic and basic media and it is also easy to prepare, which is why it is used as a catalyst for different reactions [62], and also as a support for other oxidation catalysts [63, 64].

Activated carbon can also catalyze the polymerization reactions in the presence of oxygen via oxidative coupling. Phenol oxidation over activated carbon in trickle bed reactor has been investigated [65, 66]. The activated carbon was found less active than metal oxide catalysts but more stable and more environmentally accepted, and of course cost-effective [65, 66].

Phenol conversion was compared using copper catalyst and activated carbon [67]. In the long run, copper catalyst was found to lose its activity due to leaching of copper phase. On the other hand, activated carbon also exhibited a continuous drop in phenol conversion, starting from nearly complete and finally reaching about 48%. However, the loss of activated carbon efficiency could be ascribed to its consumption during experiments; thus, the absolute activity of activated carbon remained stable during the long term.

One of the recent studies was about the CWAO of paracetamol on activated carbon [68]. The CWAO of paracetamol was investigated both as a water treatment technique and as a regenerative treatment of the carbon after adsorption in a sequential fixed bed process. They used three ACs as catalysts: a microporous basic AC and meso- and micro-porous acidic ACs. During the first CWAO experiment, they noticed that the adsorption capacity and catalytic performance of fresh basic activated carbons (S23 and C1) were higher than those of the fresh microporous acidic one (L27) despite its higher surface area. It seems that this situation changed after reuse, as finally L27 gave the best results after five CWAO cycles. Respirometric tests were also done with activated sludge and it was mentioned that in the studied conditions the use of CWAO enhanced the aerobic biodegradability of the effluent. The group also checked the different ageing by measuring the physico-chemical properties of activated carbons.

2.2.4. Application of CWAO in fine chemical industry

CWAO has been applied to many different model effluents, but relatively few studies have been devoted to real and complex industrial wastes [23, 27-28, 69-73]. Even in literature there are very limited number of works that dealt with real complex wastewaters. Our focus here is the pharmaceutical industry that produces mixtures of liquid wastes containing water and organic solvents, aside from higher molecular weight organic and inorganic compounds with different concentrations and different pH. Treating these wastewaters needs special conditions.

In a publication, the catalytic wet oxidation of wastewater originating from apramycin production was investigated with supported Ru oxide catalysts [72]. Ru catalyzed oxidation of wastewaters originating from meat processing and vegetable processing industries were also carried out [73]. Three rather detailed reviews were published concerning wet oxidation and catalytic wet oxidation [60, 74, 75]. They also mentioned the oxidation of miscellaneous organic compounds, but published no data specifically about the oxidation of pharmaceutical PWWs.

In another research, the effect of CuO/Al_2O_3 was investigated—which was prepared by consecutive impregnation—on three different azo dyes (Methyl Orange, Direct Brown, and Direct Green), which were treated by CWAO. The relationships of decolorization extent, COD, and total organic carbon (TOC) removal in the dye solution were also investigated. The 99% of color and 70% of TOC removal in 2 h indicated that the CuO/Al_2O_3 catalyst had excellent catalytic activity in treating azo dyes [76]. In Table 3, the most characteristic results of CWAO are collected; the substrates tested are in most cases phenol and acetic acid. The latter is resistant in WO, but easily biodegradable in the biological denitrification.

Catalyst		*Application*	
Active Phase	**Carrier**	**Substrate**	**Reference**
Cu	Alumina	Phenol Phenol p-cresol	Sandra et. al. 1974 Kim et al. 1991 Mishra et. al. 1993
Cu	Alumina Silica	chlorophenols	Sanger et. al. 1992
Cu-Zn-Al oxide	Alumina	phenol compounds	Pintar et al. 1992
Cu-Mg-La	Zn aluminate	acetic acid	Box et.al. 1974, Levec et. al. 1976
Mn	Alumina Sr115	phenol chlorophenols	Sandra et. al. 1974 Sanger et. al. 1992
Mn-Ce	None	polyethylene-glycol	Imamura et al.1986
Mn-Zn-Cr	None	industrial wastes	Moses et al. 1954
Cu-Co-Ti-Al	Cement	phenol	Schmidt et. al. 1990
Co	None	alcohols, amines, etc.	Ito et. al. 1989
Co-Bi	None	acetic acid	Imamura et. al.1982
Co-Ce	None	ammonia	Imamura et. al. 1985
Fe	Silica	chlorophenols	Sanger et. al. 1992
Ru	Cerium oxide	alcohols, phenols, etc	Imamura et. al. 1988
Ru-Rh	Alumina	wet oxidized sludge	Takahasi et. al. 1991
Pt-Pd	Titania-zirconia	industrial wastes	Ishii et. al. 1991
Ru	Titania-zirconia	industrial wastes sludge	Harada et. al. 1993
Pt	Alumina	phenol	Hamoudi et. al. 1998
Mn	Cerium oxide	phenol	Hamoudi et. al. 1998
Ru	Titania	phenol	Vaidya et. al. 2002
Pt, Ru	Carbon black composite Silica-titania	phenol	Cybulski et. al. 2004
Ru	Pelletized cerium oxide Zirconia	phenol	Wang et al. 2008
Ru oxide	Titania Zirconia Titania-cerium oxide Zirconia-cerium oxide	acetic acid	Wang et. al. 2008
Pt, Ru	Cerium oxide, Zr-(Ce-Pr)-O_2	phenol	Keav et. al. 2010
Ru, Ru-Ce	Alumina	isopropyl alcohol phenol acetic acid DMF	Yu et. al. 2011
Graphene oxide (GO)	None	phenol	Yang et. al. 2014
Ru, Pt	Zirconia Cerium oxide Titania	acetic acid	Lafaye et. al. 2015
Ru, Pt	Titania Cerium oxide Titania-cerium oxide	phenol	Espinosa de los Monteros et. al. 2015

Table 3. Summary of reported heterogeneous catalytic WO research [38, 57, 77-100]

2.3. Intensification of WAO with combined technologies; effect of high energy radiation on wet oxidation at elevated temperature (combination of the two methods)

As it is indicated in the previous section, one of the new areas in treating liquid wastes of high-organic content generated by the fine chemical mainly the pharmaceutical industry is the combination of AOPs with WAO. This area of work is rather new and there are very few articles dealing with this area of research.

We already know that the organic compounds undergo chain type oxidation in a reactor at high temperature and high pressure, of course sometimes in the presence of a catalyst. The high temperature is needed for initiation of oxidation processes, but at the same time this high temperature and pressure could cause serious corrosion even at alkaline pH, so it could be a good idea to decrease the temperature yet induce the chain initiation of oxidation by radiation. For this reason, WO and irradiation could be combined.

For radiation processing of polluted water, high-energy electron beam, γ -rays, or x-rays can be principally used. For producing the electron beam, electron accelerator devices are far the best radiation sources with respect to their output power and practical applicability.

2.3.1. *Water radiolysis*

In the interaction between ionizing radiation (high-energy electron beam, gamma rays) and water, electronically excited and ionized molecules are formed and the product of it will be primary species such as $\cdot OH$, e^-_{aq}, $H\cdot$, and molecular products such as H_2, H_2O_2. In the presence of oxygen in water the reducing species H-atoms and the solvated electrons (e^-_{eq}) are converted into oxidizing species, perhydroxy radicals ($HO_2\cdot$) and perhydroxyde radical anions ($HO_2\cdot$). The last one together with the $OH\cdot$ radicals are responsible for the degradation of water pollutants.

$$H_2O \rightarrow OH\bullet + e_{aq}^- + H\bullet + H_2O_2$$

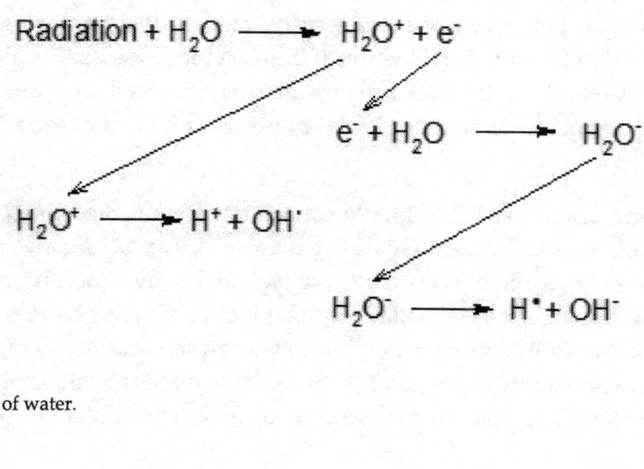

$$\text{Radiation} + H_2O \longrightarrow H_2O^+ + e^-$$

$$e^- + H_2O \longrightarrow H_2O^-$$

$$H_2O^+ \longrightarrow H^+ + OH^\cdot$$

$$H_2O^- \longrightarrow H^\cdot + OH^-$$

Figure 3. Radiolysis of water.

The radiation induced degradation of neutral phenol solution was studied in the past using end-product techniques [101-105]. There are other few works on mechanistic studies on phenol and other aromatic molecules that were carried out by combining end products and transient detections and it was suggested that the transformations were initiated by hydroxyl radical attachment to the ring and reaction of O_2 with the radicals produced [106-108].

Pulse radiolysis of 2,6-dichloroaniline in dilute aqueous solution was investigated. It is known that mono- and dichloroanilines are considered to be highly hazardous pollutants in wastewater. These compounds are important chemical intermediates of dye and plant protection agent production. In this investigation, the hydroxyl radical formed in water radiolysis was reacted with 2,6-dichloroanliline forming hydroxy-cyclohexadienyl derivative. The irradiation was carried out at room temperature by a ^{60}Co γ-source, built into a panorama irradiator, with 1.5 kGy h^{-1} dose rate. The hydroxy-cyclohexadienyl radical in the absence of dissolved O_2 partly transformed to anilino radical when oxygen was present in the reaction mixture the radical transformed to peroxy radical. According to chemical oxygen demand measurements, the reaction of one OH· radical induced the incorporation of 0.6 O_2 into the products [109].

The irradiation-induced decolorization and degradation of aqueous solutions of azo dyes and some intermediates (anilines, phenols, triazines) were successful with electron beam irradiation. The experimental methods were the pulse radiolysis and end-product analysis with HPLC-MS. Demonstrating the practical applicability of this method, a continuous irradiation device has been built. The feed was the dye containing water of red color; the effluent was the colorless liquid during contact time of less than 0.1 second with the high energy electron beam (4 MeV) generated by a linear electron accelerator [110].

The other topic is the investigation of the degradation of pharmaceuticals that can be detected in natural waters as emergent pollutants. The radiation-induced degradation of ketoprofen in dilute aqueous solution has been tested. The intermediates and final products of ketoprofen degradation were determined in 0.4 mmol/dm^3 solution by pulse radiolysis and gamma radiolysis. UV-Vis spectrophotometry and HPLC separation served for identifying the product compounds [111].

The successful degradation of organic molecules being in small concentration in waters, with high-energy irradiation generating radicals already at room temperature prompted us to combine this method with WO. Recently, this hybrid method was used and compared with the classical WO method. Noticeable conversion was observed in phenol oxidation by irradiation already at room temperature in the presence of high concentration of dissolved oxygen [112].

In one of the recent studies, WAO of highly concentrated emulsified wastewater was conducted. These kinds of wastewater usually contain all kinds of organic matters such as surfactants, additives, and mineral oils. They are typical, highly concentrated hardly biodegradable organic wastewaters. This oxidation took place in a 2 L high-pressure autoclave in batch mode. The initial COD concentration of the wastewater was 48000 mg/L. After 2 h of oxidation at 220°C with supply of oxygen 1.25 times more than its theoretical value, the COD was reduced by 86.4%. The temperature seemed to be a key influential factor, especially between 180°C and 220°C the COD and TOC removal was evidently increased. They also recognized that with increasing the initial partial pressure of oxygen (pO_2), the reaction rate significantly increased [113].

3. Conclusion

The WO of PWWs became an important disposal technology, which in combination with activated sludge treatment fulfils the strict environmental requirements as well. It works at high temperature and high pressure and often has corrosive reaction mixture; therefore, its intensification is a permanent need. The most promising solutions are the catalytic wet oxidation and the combination of WO with AOP techniques, both of generate the most reactive OH• radicals.

References

[1] Zimmermann FJ, Diddams DG. The Zimmermann process and its application in the pulp and paper industry. TAPPI. 1960;43:710-715.

[2] Mishra VS, Mahajani VV, Joshi JB. Wet air oxidation. Ind. Eng. Chem. Res. 1995;34:2-48. DOI: 10.1021/ie00040a001.

[3] Siemens AG. "Zimpro wet air oxidation systems: The cleanest way to treat the dirtiest water" brochure [Internet]. Available from: http://www.energy.siemens.com/hq/pool/hq/industries-utilities/oil-gas/portfolio/water-solution/Zimpro_Wet_Air_Oxidation.pdf [Accessed: 2015-03-05].

[4] Siemens AG, "The effects of caustic tower operations and spent caustic handling on the Zimpro wet air oxidation (WAO) of ethylene spent caustic" brochure [Internet]. Available from: http://www.energy.siemens.com/hq/pool/hq/industries-utilities/oil-gas/portfolio/water-solution/Ethylene_plant_effects_on_WAO_operations_EPC_4_2009.pdf [Accessed: 2015-03-06].

[5] Knops PCM, Bowers DL, Krielen A, Romano FJ. The start-up and operation of a commercial below-ground aqueous phase oxidation system for processing sewage sludge. VerTech Treatment Systems Technical Documentation. 1993.

[6] Li L, Chen P, Gloyna EF. Generalized kinetic model for wet oxidation of organic compounds. AIChE Journal. 1991; 37;1687. DOI: 10.1002/aic.690371112.

[7] Bachir S, Barbati S, Ambrosio M, Tordo P. Kinetics and mechanism of wet-air oxidation of nuclear-fuel-chelating compounds. Ind. Eng. Chem. Res. 2001;40:1798-1804. DOI: 10.1021/ie000818t.

[8] Robert R, Barbati S, Ricq N, Ambrosio M. Intermediates in wet oxidation of cellulose: Identification of hydroxyl radical and characterization of hydrogen peroxide. Water Res. 2002;36:4821-4829. DOI: 10.1016/S0043-1354(02)00205-1

[9] Duffy JE, Anderson MA, Hill CG, Zeltner WA. Wet peroxide oxidation of sediments contaminated with PCBs. Environ. Sci. Technol. 2000;34:3199-3204. DOI: 10.1021/es000924m.

[10] Joglekar HS, Samant SD, Joshi JB. Kinetics of wet air oxidation of phenol and substituted phenols. Water Res. 1991;25:135-145. DOI: 10.1016/0043-1354(91)90022-I.

[11] Willms RS, Balinsky AM, Reible DD, Wetzel DM, Harrison DP. Aqueous Phase Oxidation: Rate Enhancement Studies. Ind. Eng. Chem. Res. 1987;26:832-836. DOI: 10.1021/ie00063a031.

[12] Rivas FJ, Beltran FJ, Gimeno O, Acedo B. Wet air oxidation of wastewater from olive oil mills. Chem. Eng. Tech. 2001;24:415-421. DOI: 10.1002/1521-4125(200104)24:4<415::AID-CEAT415>3.0.CO;2-C.

[13] Wu Q, Hu X, Yue PL. Kinetics study on catalytic wet air oxidation of phenol. Chem. Eng. Sci. 2003;58:923-928. DOI: 10.1016/S0009-2509(02)00628-0.

[14] Sadana A, Katzer JR. Catalytic oxidation of phenol in aqueous solution over copper oxide. J. Catal., 1974;35:140-152. DOI: 10.1021/i160050a007.

[15] Rivas FJ, Kolaczkowski ST, Beltran FJ, McLurgh DB. Development of a model for the wet air oxidation of phenol based on a free radical mechanism. Chem. Eng. Sci. 1998;53:2575-2586. DOI: 10.1016/S0009-2509(98)00060-8.

[16] Bonin J, Janik I, Janik D, Bartels DM. Reaction of the hydroxyl radical with phenol in water up to supercritical conditions. J. Phys. Chem. A. 2007;111:1869-1878. DOI: 10.1021/jp0665325.

[17] Hosseini AM, Bakos V, Jobbágy A, Tardy GM, Mizsey P, Makó M, Tungler A. Cotreatment and utilization of liquid pharmaceutical wastes. Per. Pol. Chem. Eng. 2011;55:3-10. DOI: 10.3311/pp.ch.2011-1.01.

[18] Hosseini AM, Tungler A, Bakos V. Wet oxidation of process wastewaters of fine chemical and pharmaceutical origin. Reac. Kinet. Mech. Cat. 2011;103:251-260. DOI: 10.1007/s11144-011-0315-2.

[19] Hosseini AM, Tungler A, Horváth ZsE, Schay Z, Széles É. Catalytic wet oxidation of real process wastewaters. Per. Pol. Chem. Eng. 2011;55:49-57. DOI: 10.3311/pp.ch. 2011-2.02.

[20] Rivas FJ, Kolaczkowski ST, Beltran FJ, McLurgh DB. Hydrogen peroxide promoted wet air oxidation of phenol: influence of operating conditions and homogeneous metal catalysts. J. Chem. Technol. Biotechnol. 1999;74:390-398. DOI: 10.1002/(SICI)1097-4660(199905)74:5<390::AID-JCTB64>3.0.CO;2-G.

[21] Vogel F, Harf J, Hug A, von Rohr PR. Promoted Oxidation of phenol in aqueous solution using molecular oxygen at mild conditions. Env. Prog. 1999;18:7-13. DOI: 10.1002/ep.670180114.

[22] Devlin HR, Harris IJ. Mechanism of the oxidation of aqueous phenol with dissolved oxygen. Ind. Eng. Chem. Fundam. 1984;23:387-392. DOI: 10.1021/i100016a002.

[23] Pintar A, Gorazol B, Besson M, Gallezot P. Catalytic wet-air oxidation of industrial effluents: total mineralization of organics and lumped kinetic modeling. Appl. Cat. B. Env. 2004;47:143-152. DOI: 10.1016/j.apcatb.2003.08.005.

[24] Suarez-Ojeda ME, Carrera J, Metcalfe IS, Font J. Wet Air oxidation (WAO) as a precursor to biological treatment of substituted phenols: Refractory Nature of the WAO intermediates. Chem. Eng. J. 2008;144:205-212. DOI:10.1016/j.cej.2008.01.022.

[25] Barbier J Jr, Delanoë F, Jabouille F, Duprez D, Blanchard G, Isnard P. Total oxidation of acetic acid in aqueous solutions over noble metal catalysts. J. Catal. 1998;177:378-385. DOI: 10.1006/jcat.1998.2113.

[26] Imamura S. Wet oxidation of a model domestic wastewater on a Ru/Mn/Ce composite catalyst, Ind. Eng. Chem. Res. 1999;38:1743-1753. DOI: 10.1021/ie970538m.

[27] Luck F. Wet air oxidation: past, present and future. Catal. Today. 1999;53:81-91. DOI: 10.1016/S0920-5861(99)00112-1.

[28] Matatov-Meytal YI, Sheintuch M. Catalytic Abatement of Water Pollutants. Ind. Eng. Chem. Res. 1998;37:309-326. DOI: 10.1021/ie9702439.

[29] Fortuny A, Font J, Fabregat A. Wet air oxidation of phenol using active carbon as catalyst. Appl. Cat. B. 1998;19:165-173. DOI: 10.1016/S0926-3373(98)00072-1.

[30] Pintar A, Batista J, Tisler T. Catalytic wet-air oxidation of aqueous solutions of formic acid, acetic acid and phenol in a continuous-flow trickle-bed reactor over Ru/TiO_2 catalysts. Appl. Cat. B. Env. 2008;84:30-41. DOI: 10.1016/j.apcatb.2008.03.001.

[31] Mantzavinos D, Hellenbrand R, Livingston AG, Metcalfe IS. Catalytic wet air oxidation of polyethylene glycol. Appl. Cat. B. Env. 1996;11:99-119. DOI: 10.1016/S0926-3373(96)00037-9.

[32] Mikulová J, Rossignol S, Barbier J Jr, Mesnard D, Kappenstein C, Duprez D. Ruthenium and platinum catalysts supported on Ce, Zr, Pr-O mixed oxides prepared by soft chemistry for acetic acid wet air oxidation. Appl. Cat. B. Env. 2007;72:1-10. DOI: 10.1016/j.apcatb.2006.10.002.

[33] Wu Q, Hu X, Yue P. Kinetics study on catalytic wet air oxidation of phenol. Chem. Eng. Sci. 2003;58:923-928. DOI: 10.1016/S0009-2509(02)00628-0.

[34] Fortuny A, Ferrer C, Bengoa C, Font J, Fabregat A. Catalytic removal of phenol from aqueous phase using oxygen or air as oxidant. 1995;24:79-83. Catal. Today. DOI: 10.1016/0920-5861(95)00002-W.

[35] Doornkamp C, Ponec V. The universal character of the Mars and Van Krevelen mechanism. J. Mol. Catal. A. 2000;162:19-32. DOI: 10.1016/S1381-1169(00)00319-8.

[36] Mars P, van Krevelen DW. Oxidations carried out by means of vanadium oxide catalysts. Spec. Suppl. Chem. Eng. Sci. 1954;8:41-59. DOI: 10.1016/S0009-2509(54)80005-4.

[37] Chauzy J, Martin JC, Cretenot D, Rosiere JP. Wet air oxidation of municipal sludge: Return experience of the North Brussels Waste Water Treatment Plant. Water Practice & Technology. 2010;5;1. DOI:10.2166/wpt.2010.003.

[38] Ishii T, Mitsui K, Sano K, Inoue A, Method for treatment of wastewater. US Patent No. 5145587. 1991.

[39] Cooke BH, Taylor MR. The environmental benefit of coal gasification using the BGL gasifier. Fuel. 1993;72:305-314. DOI: 10.1016/0016-2361(93)90047-6.

[40] Pintar A, Levec J. Catalytic liquid-phase oxidation of phenol aqueous solutions: A kinetic investigation. Ind. Eng. Chem. Res. 1994;33:3070-3077. DOI: 10.1021/ie00036a023.

[41] Duprez D, Delanoë JF, Barbier J Jr, Isnard P, Blanchard G. Catalytic oxidation of organic compounds in aqueous media. Catal. Today. 1996;29:317-322. DOI: 10.1016/0920-5861(95)00298-7.

[42] Gallezot P, Laurain N, Isnard P. Catalytic wet-air oxidation of carboxylic acids on carbon-supported platinum catalysts. Appl. Catal. B. Env. 1996;9:11-17. DOI: 10.1016/0926-3373(96)90070-3.

[43] Gallezot P, Chaumet S, Perrard A, Isnard P. Catalytic wet air oxidation of acetic acid on carbon-supported ruthenium catalysts J. Catal. 1997;168:104-109. DOI:10.1006/jcat.1997.1633.

[44] Imamura S, Fukuda I, Ishida S. Wet oxidation catalyzed by ruthenium supported on cerium (IV) oxides. Ind. Eng. Res. 1988;27:718-721. DOI: 10.1021/ie00076a033.

[45] Mikulová J, Rossignol S, Barbier J Jr, Duprez D, Kappenstein C. Characterizations of platinum catalysts supported on Ce, Zr, Pr-oxides and formation of carbonate species in catalytic wet air oxidation of acetic acid. Catal. Today. 2007;124:185-190. DOI: 10.1016/j.cattod.2007.03.036

[46] Wang J, Zhu W, He X, Yang S. Catalytic wet air oxidation of acetic acid over different ruthenium catalysts. Catal. Commun. 2008;9:2163-2167. DOI: 10.1016/j.catcom.2008.04.019.

[47] Fortuny A, Miro C, Font J, Fabregat A. Water pollution abatement by catalytic wet air oxidation in a trickle bed reactor. Catal. Today, 1999;53:107-114. DOI: 10.1016/S0920-5861(99)00106-6.

[48] Akyurtlu JF, Kovenklioglu S, Akyurtlu A. Catalytic oxidation of phenol in aqueous solution. Catal. Today. 1998;40:343-352. DOI: 10.1016/S0920-5861(98)00063-7.

[49] Pintar A, Levec J. Catalytic oxidation of aqueous p-chlorophenol and p-nitrophenol solutions. Chem. Eng. Sci. 1994;49:4391-4407. DOI: 10.1016/S0009-2509(05)80029-6.

[50] Qin J, Zhang Q, Chuang KT. Catalytic wet oxidation of p-chlorophenol over supported noble metal catalysts. Appl. Catal. B. Env. 2001;29:115-123. DOI: 10.1016/S0926-3373(00)00200-9.

[51] Huang TL, Macinnes JM, Cliffe KR. Nitrogen removal from wastewater by a catalytic oxidation method. Water Res. 2001;35:2113-2120. DOI: 10.1016/S0043-1354(00)00492-9.

[52] Pereira MFR, Orfao JJM, Figueiredo JL, Oxidative dehydrogenation of ethylbenzene on activated carbon catalysts: 2. Kinetic modeling. Appl. Catal. A. Gen. 2000;196:43-54. DOI: 10.1016/S0926-860X(99)00447-0.

[53] Oliviero L, Wahyu H, Barbier J Jr, Duprez D, Ponton JW, Metcalfe IS, Mantzavinos D. Experimental and predictive approach for determining wet air oxidation reaction pathways in synthetic wastewaters. Chem. Eng. Res. Des. 2003;81:384-392. DOI: 10.1205/026387603605969372.

[54] Akolekar DB, Bhargava SK, Shirgoankar I, Prasad J. Catalytic wet oxidation: An environmental solution for organic pollutant removal from paper and pulp industrial waste liquor. Appl. Catal. A. Gen. 2002;236:255-262. DOI: 10.1016/S0926-860X(02)00292-2.

[55] Lee DK, Kim DS. Catalytic wet air oxidation of carboxylic acids at atmospheric pressure. Catal. Today. 2000;63:249-255, DOI: 10.1016/S0920-5861(00)00466-1.

[56] Pintar A, Besson M, Gallezot P. Catalytic wet air oxidation of Kraft bleach plant effluents in a trickle-bed reactor over a Ru/TiO2 catalyst. Appl. Catal. B. Env. 2001;31:275-290. DOI: 10.1016/S0926-3373(00)00288-5.

[57] Imamura S, Nakamura M, Kawabata N, Yoshida J, Ishida S. Wet oxidation of poly(ethylene glycol) catalyzed by manganese-cerium composite oxide, Ind. Eng. Chem. Prod. Res. Dev. 1986;25:34-37. DOI: 10.1021/i300021a009.

[58] Pirkanniemi K, Sillanpaa M. Heterogeneous water phase catalysis as an environmental application: a review. Chemosphere. 2002;48:1047-1060, DOI: 10.1016/S0045-6535(02)00168-6.

[59] Levec J, Pintar A. Catalytic wet-air oxidation processes: A review. Catal. Today. 2007;124:172-184. DOI: 10.1016/j.cattod.2007.03.035.

[60] Singh B, Madhusudhanau S, Bubey V, Nath R, Rao NBSN. Active carbon for removal of toxic chemicals from contaminated water. Carbon. 1996;34:327-330. DOI: 10.1016/0008-6223(95)00179-4.

[61] Hu X, Lei L, Chu HP, Po LY. Copper/Activated carbon as catalyst for organic wastewater treatment. Carbon. 1999;37:631-637, DOI: 10.1016/S0008-6223(98)00235-8.

[62] Coughlin RW. Carbon as adsorbent and catalyst. Ind. Eng. Chem. Prod. Res. Dev. 1969;8:12-23. DOI: 10.1021/i360029a003.

[63] Aksoylu AE, Madalena M, Freitas A, Pereira MFR, Figueiredo JL. The effects of different activated carbon supports and support modifications on the properties of Pt/AC catalysts. Carbon. 2001;39:175-185. DOI: 10.1016/S0008-6223(00)00102-0.

[64] Birbara PJ, Genovese JE. US Patent No 5 362 405 (1995).

[65] Tukac V, Hanika J. Catalytic wet oxidation of substituted phenols in the trickle bed reactor, J. Chem. Technol. Biotechnol. 1998;71:262-266. DOI: 10.1002/(SICI)1097-4660(199803)71:3<262::AID-JCTB855>3.0.CO;2-R.

[66] Tukac V, Vokal J, Hanika J. Mass transfer-limited wet oxidation of phenol. J. Chem. Technol. Biotechnol. 2001;76:506-510. DOI: 10.1002/jctb.402.

[67] Fortuny A, Font J, Fabregat A. Wet air oxidation of phenol using active carbon as catalyst. Appl. Catal. B. Env. 1998;19:165-173. DOI: 10.1016/S0926-3373(98)00072-1.

[68] Quesada PI, Julcour-Lebigue C, Jauregui-Haza UJ, Wilhelm AM, Delmas H. Degradation of paracetamol by catalytic wet air oxidation and sequential adsorption - Catalytic wet air oxidation on activated carbons. J. Haz. Mat. 2012;221-222:131-138. DOI: 10.1016/j.jhazmat.2012.04.021.

[69] Suarez-Ojeda EM, Guisalsola A, Baeza JA, Fabregat A, Stuber F, Fortuny A, Carrera J. Integrated catalytic wet air oxidation and aerobic biological treatment in a municipal WWTP of high strength o-cresol waste water. Chemosphere. 2007;66:2096-2105. DOI: 10.1016/j.chemosphere.2006.09.035.

[70] Imamura S. Catalytic and noncatalytic wet oxidation. Ind. Eng. Chem. Res. 1999;38:1743-1753. DOI: 10.1021/ie980576l

[71] Lei YJ, Zhang SD, He JC, Wu JC, Yang Y. Ruthenium catalyst for treatment of water containing concentrated organic waste. Platinum Met. Rev. 2005;49:91-97. DOI: 10.1595/147106705x45640.

[72] Yang SX, Feng YJ, Wan FJ, Lin QY, Zhu WP, Jiang ZP. Pretreatment of apramycin wastewater by catalytic wet air oxidation. J. Env. Sci. 2005;17:623-626. DOI: 1001-0742(2005)04-0623-04.

[73] Heponiemi A, Rahikka L, Lassi U, Kuokkanen T. Catalytic oxidation of industrial wastewaters a comparison study using different analyzing methods. In: Chemical Engineering Transactions, 9th International Conference on Chemical and Process Engineering; 10-13 May 2009; Rome, Italy.

[74] Bhargava SK, Tardio J, Prasad J, Föger K, Akolekar DB, Grocott SC. Wet oxidation and catalytic wet oxidation. Ind. Eng. Chem. Res. 2006;45:1221-1258. DOI: 10.1021/ie051059n.

[75] Cybulski A. Catalytic wet air oxidation: are monolithic catalysts and reactors feasible? Ind. Eng. Chem. Res. 2007;46:4007-4033. DOI: 10.1021/ie060906z.

[76] Hua L. Degradation process analysis of the azo dyes by catalytic wet air oxidation with catalyst CuO/γ-Al2O3. Chemosphere. 2013;90:143-149. DOI: 10.1016/j.chemosphere.2012.06.018.

[77] Sadana A, Katzer JR. Catalytic oxidation of phenol in aqueous solution over copper oxide. Ind. Eng. Chem. Fundam. 1974;13:127-134. DOI: 10.1021/i160050a007.

[78] Kim S, Shah YT, Cerro RL, Abraham MA. Aqueous phase oxidation of phenol in a monolithic reactor. In: Proceedings of AIChE Annual Meeting; Miami Beach, Florida, 1992.

[79] Mishra VS, Joshi JB, Mahajani VV, Kinetics of p-cresol destruction via wet air oxidation. Indian Chemical Engineer 1993;35;211.

[80] Smith KJ, Sanford EC, editors. Sanger AR, Lee TTK, Chuang KT. Cu/Al2O3, silica, Fe/ silica for clorophenol treatment. Progress in Catalysis. Elsevier, 1992; 197.

[81] Pintar A, Levec J. Catalytic liquid-phase oxidation of refractory organics in waste water. Chem. Eng. Sci. 1992;47:2395-2400. DOI: 10.1016/0009-2509(92)87066-Y.

[82] Box E. O., Farha F. Polluted water purification. U.S. Patent No. 3823088. 1974.

[83] Levec J, Herskowitz M, Smith JM. An active catalyst for the oxidation of acetic acid solutions. AlChe. J. 1976;22:919-920. DOI: 10.1002/aic.690220516.

[84] Moses DV, Smith EA. A process for the disposal of industrial wastes of an organic nature by contact with a solid catalyst of manganese-zinc-chromium at a temperature of 100 U.S. Patent No. 2690425. 1954.

[85] Schmidt FK, Kochetkova RP, Babikov AF, Shiverskaia IP, Shpilevskaia LI, Eppel CA. In: Proceedings of the 8th French-Soviet Meeting on Catalysis; June 1990; Novossibirsk, Russia. p.140.

[86] Ito MM, Akita K, Inoue H. Wet oxidation of oxygen and nitrogen containing compounds catalyzed by cobalt (III) oxide. Ind. Eng. Chem. Res. 1989;28:894-899. DOI: 10.1021/ie00091a003.

[87] Imamura S, Hirano A, Kawabata N. Wet oxidation of acetic acid catalyzed by Co-Bi complex oxides, Ind. Eng. Chem. Prod. Res. Dev. 1982;21:570-575. DOI: 10.1021/i300008a011.

[88] Imamura S, Doi A, Ishida S. Wet oxidation of ammonia catalyzed by cerium based composite oxides. Ind. Eng. Chem. Prod. Res. Dev. 1985;24:75-80. DOI: 10.1021/i300017a014.

[89] Imamura S, Fukuda I, Ishida S. Wet oxidation catalyzed by ruthenium supported on cerium (IV) oxides. Ind. Eng. Chem. Res. 1988;27:718-721. DOI: 10.1021/ie00076a033.

[90] Takahasi Y, Takeda N, Aoyagi A, Tanaka K. Catalytic wet-oxidation of human waste produced in a space habitat: purification of the oxidized liquor for human drinking. In: Proceedings of 4th European Symposium on Space Environmental Control Systems; 21-24 October 1991; Florence, Italy; p. 643-648.

[91] Harada Y, Yamasaki K. Treatment of wastewater and sludge by a catalytic wet oxidation process. In: Proceedings of IDA/WRPC World Conference on Desalination Water Treatment; 3-6 November 1993; Yokohama, Japan; p. 231-242..

[92] Hamoudi S, Larachi F, Sayari A. Wet oxidation of phenolic solutions over heterogeneous catalysts: degradation profile and catalyst behavior. J. Catal. 1998;177;247-258 DOI: 10.1006/jcat.1998.2125.

[93] Vaidya PD, Mahajani VV. Insight into heterogeneous catalytic wet oxidation of phenol over a Ru/TiO2 catalyst. Chem. Eng. Journ. 2002;87;403-416 DOI: 10.1016/S1385-8947(02)00020-7.

[94] Cybulski A, Trawczynski J. Catalytic wet air oxidation of phenol over platinum and ruthenium catalysts. Appl. Cat. B. Env. 2004;47;1-13 DOI: 10.1016/S0926-3373(03)00327-8.

[95] Wang J, Zhu W, Yang S, Wang W, Zhou Y. Catalytic wet air oxidation of phenol with pelletized ruthenium catalysts. Appl. Cat. B. Env. 2008;78;30-37 DOI: 10.1016/j.apcatb.2007.08.014.

[96] Keav S, Martin A, Barbier Jr. J, Duprez D. Deactivation and reactivation of noble metal catalysts tested in the catalytic wet air oxidation of phenol. Catal. Today. 2010;151;143-147 DOI: 10.1016/j.cattod.2010.01.025.

[97] Yu C, Zhao P, Chen G, Hu B. Al_2O_3 supported Ru catalysts prepared by thermolysis of $Ru_3(CO)_{12}$ for catalytic wet air oxidation. Appl. Surf. Sci. 2011;257;7727-7731 DOI: 10.1016/j.apsusc.2011.04.017.

[98] Yang S, Cui Y, Sun Y, Yang H. Graphene oxide as an effective catalyst for wet air oxidation of phenol. J. Haz. Mat. 2014;280;55-62 DOI: 10.1016/j.jhazmat.2014.07.051.

[99] Lafaye G, Barbier Jr. J, Duprez D. Impact of cerium-based support oxides in catalytic wet air oxidation:Conflicting role of redox and acid-base properties. Catal. Today. 2015. DOI: 10.1016/j.cattod.2015.01.037.

[100] Monteros AE dl, Lafaye G, Cervantes A, Del Angel G, Barbier Jr. J. Catalytic wet air oxidation of phenol over metal catalyst (Ru,Pt)supported on TiO_2-CeO_2 oxides. Catal. Today. 2015. DOI: 10.1016/j.cattod.2015.01.009.

[101] Hashimoto S, Miyata T, Washino M, Kawakami W. Decoloration and degradation of an anthraquinone dye aqueous solution in flow system using an electron beam accelerator. Radiat. Phys. Chem. 1979;13:107-113. DOI: 10.1016/0146-5724(79)90057-8.

[102] Hashimoto S, Miyata T, Suzuki N, Kawakami W. A liquid chromatographic, study on the radiolysis of phenol in aqueous solution. Environ. Sci.Technol. 1979;13:71-75. DOI: 10.1021/es60149a008.

[103] Hashimoto S, Miyata T, Kawakami W. Radiation-induced decomposition of phenol in flow system. Radiat. Phys. Chem. 1980;16:59-65. DOI: 10.1016/0146-5724(80)90114-4.

[104] Sato K, Takimoto K, Tsuda S. Degradation of aqueous phenol solution by gamma irradiation. Environ. Sci. Technol. 1978;12:1043-1046. DOI: 10.1021/es60145a016.

[105] Bonin, J., Janik, I., Janik, D., Bartels, D.M., Reaction of hydroxyl radical with phenol in water up to supercritical conditions. J. Phys. Chem. A 2007;111:1869-1878. DOI: 10.1021/jp0665325.

[106] von Sonntag C, Schuchmann HP. The chemistry behind the application of ionizing radiation in water-pollution abatement. Stud. Phys. Theor. Chem. 2001;87:657-670. DOI: 10.1016/S0167-6881(01)80025-2.

[107] von Sonntag C. Free-radical-induced DNA damage and its repair: A chemical perspective. Springer Berlin Heidelberg. 2006. DOI: 10.1007/3-540-30592-0.

[108] Kozmér Zs, Arany E, Alapi T, Takács E, Wojnárovits L, Dombi A. Determination of the rate constant of hydroperoxyl radical reaction with phenol. Rad. Phys. Chem. 2014;102:135-138. DOI: 10.1016/j.radphyschem.2014.04.029.

[109] Homlok R, Takács E, Wojnárovits L. Ionizing radiation induced reactions of 2,6-dichloroaniline in dilute aqueous solution. Rad. Phys. Chem. 2012;81:1499-1502. DOI: 10.1016/j.radphyschem.2011.11.017.

[110] Wojnárovits L, Takács E. Irradiation treatment of azo dye containing wastewater: An overview. Rad. Phys. Chem. 2008;75:225-244. DOI: 10.1016/j.radphyschem. 2007.05.003.

[111] Illés E, Takács E, Dombi A, Gajda-Schrantz K, Gonter K, Wojnárovits L. Radiation induced degradation of ketoprofen in dilute aqueous solution. Rad. Phys. Chem. 2012;81:1479-1483. DOI: 10.1016/j.radphyschem.2011.11.038.

[112] Chamam M, Földváry MCs, Hosseini AM, Tungler A, Takács E, Wojnárovits L. Mineralization of aqueous phenolate solutions: A combination of irradiation treatment and wet oxidation. Rad. Phys. Chem. 2012;81;1484-1488. DOI: 10.1016/j.radphyschem. 2011.11.013.

[113] Tang W, Zeng X, Zhao J,Gu G, Li Y, Ni Y. The study on the wet air oxidation of highly concentrated emulsified wastewater and its kinetics. Sep. Purif. Technol. 2003;31:77-82. DOI: 10.1016/S1383-5866(02)00161-2.

Chapter 7

Modeling Wastewater Evolution and Management Options under Variable Land Use Scenarios

Abstract

The development of a reliable decision support system and predictions for water quantity and quality often require a reasonable level of environmental and hydrological simulations at various geographic scales. The Soil and Water Assessment Tool (SWAT) model offers distributed parameter and continuous time simulation, and flexible watershed configuration and with the adoption of geographic information system (GIS) technology, a user-friendly and interactive decision support system can be developed for wastewater management. In this chapter, we evaluated the spatio-temporal evolution of wastewater contaminants in an environmentally degraded watershed through integrated field-based investigations and modeling approach. Later, management options were identified to improve the watershed health and agro-environment. The results of the modeling study exhibited variable responses of surface runoff and water quality to different scenarios of land use change. Temporal wastewater analysis indicated a significant impact of seasonality on the contaminants' population levels. The adopted approach would prove effective in evaluating better management options to reduce negative impacts of wastewater and contaminants for sustainable agro-environment in future.

Keywords: Wastewater modeling, land use scenarios, water policy

1. Introduction

Water is a scarce source in arid and semiarid areas where most of the countries face pressure due to limited opportunities to explore new water resources. This necessitates that all potential unutilized resources of water be used to increase agriculture production. The changes in surface and subsurface flows and land use conditions have direct affect on the downstream in the form of floods and/or water quality deterioration. Climate change and human interference could lead to significant spatio-temporal variations of water quantity, quality, and the associated ecological conditions besides affecting the related management systems [1]. Such complexities force researchers to develop more robust mathematical methods and tools to analyze the relevant information, simulate the related processes, assess the potential impacts/risks, and generate sound decision alternatives. Spatially meaningful simulation of environmental flows and storages at the catchment scale is essential for predicting water quantity and quality, as well as operational management of the system [2]. There are numerous modeling wastewater efforts undertaken globally by different researchers (e.g., [1, 3–5]), the ultimate focus of which is mainly to mitigate sediment, contaminants, and non-point source nutrient; enhance water quality; and improve sustainability in agricultural production by increasing resilience. Unforeseen and undesirable consequences can result if biophysical and human systems are not examined together [6, 7]. Daloğlu et al. [4] presented a modeling framework that synthesizes social, economic, and ecological aspects of landscape change to evaluate how different agricultural policy and land tenure scenarios and land management preferences affect landscape pattern and downstream water quality. Wrede et al. [3] evaluated the performance of a fully distributed conceptual hydrologic model based on the Hydrologiska Byråns Vattenbalansavdelning (HBV) and Tracer Aided Catchment model-Distributed (TAC^D) model concepts in the Central Swedish lowlands. Nesmerak and Blazkova [8] employed a simple transfer function (SISO model) to describe the relationship between the daily total precipitation and the wastewater discharge at the inflow to the wastewater treatment plant (WWTP) for a large city. However, scientific quantifications were required on temporal and spatial scale to identify any feasible wastewater management solution rather than spot and one time sampling of effluents as reported by several studies (e.g., see [9–11]).

The development of a sufficient understanding on which to base decisions or make predictions often requires consideration of a multitude of data of different types and with varying levels of uncertainty [12]. Wastewater contains chemicals such as nitrogen, phosphorus and levels of dissolved oxygen, as well as others that may affect its composition and pH rating. Agricultural runoff, drainage, as well as inputs from municipal and industrial wastewater often degrade the quantity and quality of surface water bodies. There is a serious need for appropriate water quality monitoring for future planning and management of clean water resources. The SWAT model offers distributed parameter and continuous time simulation, and flexible watershed configuration and with the adoption of GIS technology, a user-friendly and interactive decision support system can be developed for wastewater management. The primary focus of this chapter is to assess the spatio-temporal evolution of wastewater contaminants through the modeling approach and identify management options to improve the watershed health and agro-environment. The findings of the study may support policy makers, researchers, and water managers to make more robust water policy and management options under the changing environment in the future.

2. Case study

In order to develop suitable watershed management strategies, reliable investigation of the watershed problems is necessary. The influence of historical land use evolution on the yield of Rawal watershed lying in the southern Himalayan region was studied to take proactive measures to control the negative impacts of water contaminants in the downstream. The runoff from heavy rains brings a lot of sediments and wastewater from the adjoining areas that increases suspended, as well as bed load in the Korang River and ultimately in the Rawal lake. When the organic nutrients are added to the lake it causes eutrophication—as algal growth increases in the waste, dissolved oxygen concentrations are depleted and increase in sedimentation deteriorates water quality [13]. If wastewater is being discharged into the lake, then the nutrients that are of most important concern are nitrogen and phosphorus. Different studies on the Rawal watershed revealed water quality implications at the lake site. For example, biochemical oxygen demand (BOD) of about 680 mgl^{-1} was reported by Malik [9], while Ahmad et al. [11] reported total dissolve solids (TDS) of dam water from 131mgl^{-1} to 182mgl^{-1}. The issue of dam water quality is further being aggravated by rapid unplanned urban encroachment in the Rawal watershed area since the last decade. These urban settlements are producing sewage in large quantities that ultimately drain toward the lake through freshwater streams. Since dam water supplies were mainly being used for drinking purpose, therefore, scientific investigation was required to trace the impact of urbanization on Rawal watershed runoff. Because of continuous water quantity and quality degradations primarily due to urban encroachments within the Rawal watershed, there was pressing need for its management on sustainable basis. The emphasis was to mitigate negative water quality implications particularly due to urban sewage. However, scientific quantification of waste flows over a reasonable time frame was essentially required for effective mitigation.

Initially a questionnaire-based detailed survey was undertaken by selecting Bharakaho as a pilot area with specific research motivations, e.g., to investigate water consumption and wastewater disposal systems of selected urban settlement of the Rawal watershed, to quantify temporal and spatial wastewater quality and quantity of selected locations of the settlement using statistical means, and to develop recommendations for the appropriate measures of safe and cost-effective disposal of wastewater. An existing fully calibrated and verified SWAT model of the Rawal watershed [14] was used to simulate nutrient load responses under variable land use scenarios. The model calibration was performed using daily-observed flow data for a 20-year period (2001–2010). Remote sensing (RS) image data was used to analyze the agriculture and urban land use conditions and input to the model for simulating water quality parameters, e.g., sediment, nitrogen, and phosphorus loss from the landscape.

2.1. Water quality model – SWAT

The biophysical water quality model—SWAT is a distributed model that integrates land management decisions with soil properties, climate information, and land topography to estimate water quality metrics at the watershed or river basin scale [15]. It is widely used for evaluating and predicting the impacts of conservation practices through simulating the effects of climate and land use changes on nutrient and sediment delivery from watersheds [16]. This process-based model (covering multi aspects of hydrology, soil, crop growth, nutrients, sedimentation, pesticides) divides watersheds into sub-basins and hydrologic response units (HRU) as its fundamental computational unit. Runoff flow, sediment, and nutrient loads are calculated separately for each HRU and then summed to determine the total load contribution from each sub-basin [17]. Land management decisions are represented at the HRU scale [4].

Daloğlu et al. [4] studied the impact of plausible future policy and land tenure scenarios on the delivery of available dissolved reactive phosphorus (DRP) and total phosphorus (TP) by exploring links between human and environmental systems. High surface water concentrations of nitrogen and phosphorus are correlated with inputs from fertilizers used for crops [18–20].

2.2. Description of the study area

The study area is the watershed of Rawal dam that caters water requirements of the twin cities of Islamabad and Rawalpindi, located in the Northern half of Pakistan (Figure 1). The Rawal watershed has been stretched over an area of about 272 sq km within longitudes 73º 03′ -73º 24′ E and latitudes 33° 41′ -33° 54′ N. The area falls under the scrub forest zone and supports mixture composed of *Olea ferogenia* (Wild Olive), *Dodonea viscosa, Crissa spinarum, Acacia modesta* [21]. In addition, there exist different grass species in which relative cover of *Themeda anathera* is maximum. It has been estimated that during an average year, the Rawal watershed area was draining about 84,000 acre-feet of runoff water through four major and 43 minor stream networks [22]. The Rawal dam was constructed over Korang river during 1960 at the toe of Rawal watershed area in Islamabad to harvest runoff water to primarily meet the drinking water requirements of the twin cities. However, land use changes within the Rawal watershed at the cost of deforestation have already affected the storage capacity of the dam. Rapid unplanned urbanization particularly in the lower valleys of the Rawal watershed over the last many years has emerged as major sustainability threat for the dam due to the continuously deteriorating water quality.

According to Ghumman [23], human settlements, deforestation, pesticides, erosion, and wastes from poultry, agricultural activities, and recreational activities are the most possible reasons of contamination of the water of Rawal Lake. Untreated effluents from communal, agricultural, and poultry sectors are seriously damaging the water quality of the lake. In addition to the pollution generated by human activity, the lake also receives natural pollutants that contain the excreta of various wild animal species and fouls that enter the lake via heavy rainfall [24]. Bacteria decompose this organic matter in the presence of oxygen, thus oxygen depletion results in the Eutrophication of the lake. Similarly significant eutrophication is caused by agricultural runoff, concomitant soil erosion, and point-source discharges [25]. The land use patterns within the Rawal watershed have been changed significantly since the 1960's and major catchment area has been deforested to accommodate the rapidly increasing urban population of Islamabad—the capital city of Pakistan—and other infrastructural developments. During 18-year period (from 1992 to 2010), about 53% of Rawal watershed land use has been changed [26]. The changed land use features altered watershed hydrology and conse-

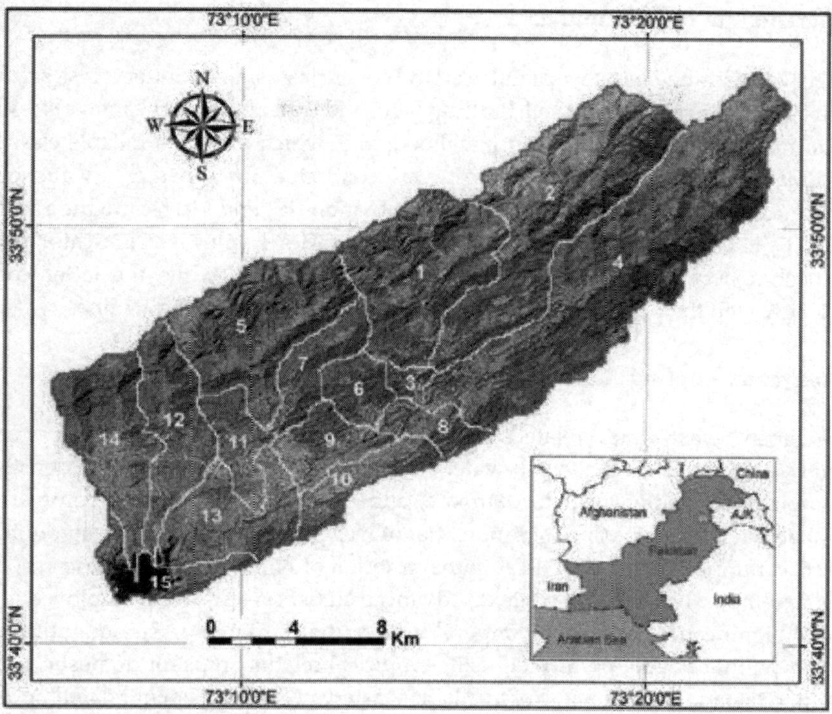

Figure 1. Location of Rawal watershed and its sub-basins.

quently the designed storage capacity of Rawal dam (47,500 acre -feet at the time of construction) has been decreased to around 31,000 acre feet [27]. Moreover, the increased urbanization, commercial, and agricultural developments within the Rawal watershed area has also significantly deteriorated the water quality [24]. Agricultural runoff results in algal blooms, poor water clarity, and summer hypoxia (low oxygen) [28–30] that generally impact fisheries, recreation, and drinking water [31].

Bharakaho, located at about 5 km from the dam site within the Rawal watershed, was selected as a surveyed site for the present study. It is the largest urban setting in the watershed. The study sub-watershed comprises of five catchments that ultimately drain into Korang river. The selected catchments were: i) Shahdara catchment (before the bridge on Murree road), ii) Colonel Amanullah road catchment, iii) Hathala catchment, IV) Kiani road catchment and Shahdara catchment (After bridge on Murree road).

3. Material and methods

In order to evaluate water quality response to varying land use changes in the Rawal watershed through hydrological modeling, satellite remote sensing data of Landsat ETM+ (Enhanced Thematic Mapper plus) of 2010 period was used in the present study.

3.1. Preparation for model input data

The image was classified into seven land use/land cover classes, e.g., conifer forest, scrub forest, agriculture land, rangeland, bare soil, built-up land, and water bodies. The image classification was undertaken using the maximum likelihood rule, which provides reliable classification results [26]. Rainfall-runoff model SWAT was calibrated for the target watershed. The statistical measures, such as coefficient of determination (R^2) and Nash-Suttchiffe Simulation Efficiency (E_{ws}), were used to evaluate model prediction. The R^2 value is an indicator of strength of relationship between the observed and simulated values, while the latter coefficient indicates how well the plot of observed versus simulated value fits the 1:1 line.

3.2. Wastewater sampling strategy

To complement wastewater related quantifications, a wastewater sampling study was designed by selecting Bharakaho sub-watershed due to its highest population density; rapid urbanization rate, location near to dam site, and existence of significant commercial and industrial activities, thus posing high potential of wastewater yield. Wastewater quality was monitored continuously for over 14 months in terms of eight parameters and results were analyzed in temporal and spatial context. Advanced statistical tools were employed to further investigate significance levels for temporal and spatial variability. The quantifications of wastewater pollution were performed using empirical relationships for domestic, industrial, and commercial land uses. Review of available wastewater management related options was made and based on a robustly developed criterion, the feasible option was recommended. Five catchments and their drainage pattern were delineated in Bharakaho sub-watershed using GIS tools. The sampling points were selected based upon physical surveys at outlets of these catchments (Figure 2), the particulars of which are shown in Table 1.

Location	Geographic Coordinates	Area (ha)	Total Industrial and Commercial Activities
Shahdara catchment (Before bridge on Murree road)	73° 10' 23.7128" E 33° 44' 12.0721" N	132	3
Col. Amanullah Road catchment	73° 10' 23.7128" E 33° 44' 13.0822" N	25	7
Hathala catchment	73° 11' 38.9016" E 33° 44' 38.2142" N	169	20
Kiani road catchment	73° 10' 42.1186" E 33° 43' 35.1740" N	189	19
Shahdara catchment (After bridge on Murree road)	73° 10' 37.0289" E 33° 43' 29.6024" N	160	11

Table 1. Location of sampling points and type of activities in the study area.

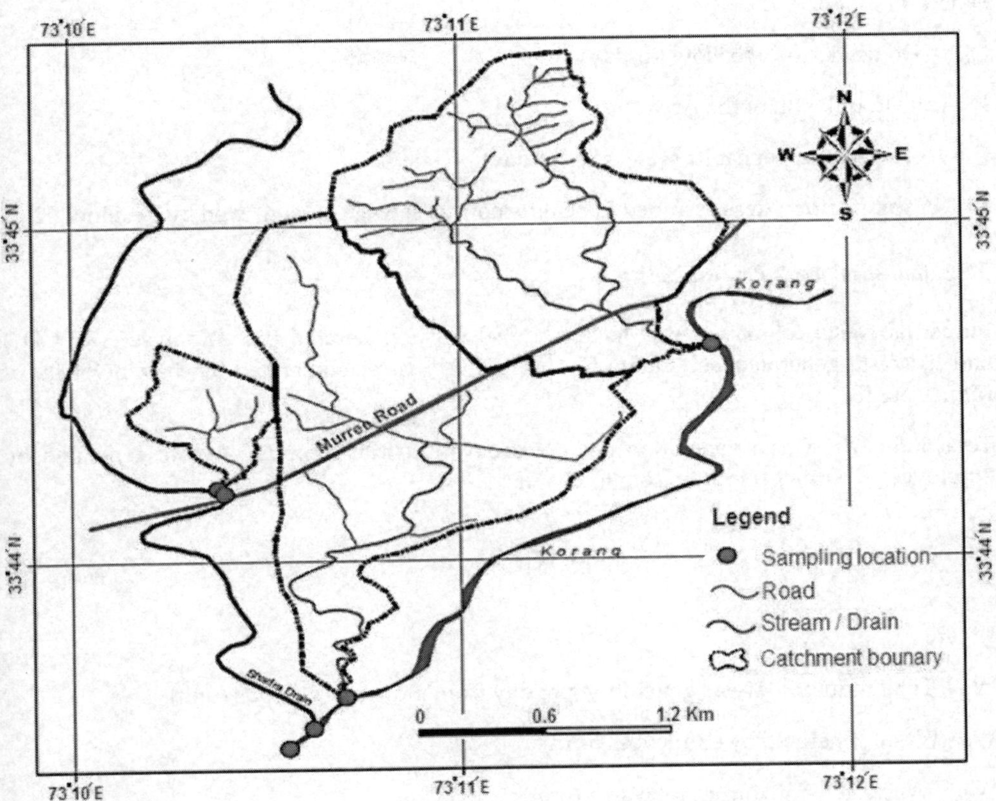

Figure 2. Location of sampling points in Bharakaho sub-watershed.

To analyze wastewater implications, extensive review was made. The critical parameters related wastewaters associated with urbanization selected for monitoring purposes are: Biochemical Oxygen Demand (BOD); Chemical Oxygen Demand (COD); Electrical Conductivity (EC); pH; Total Dissolve Solids (TDS); Total Phosphorous (TP); Nitrate (NO_2); and Nitrite (NO_3). Monthly sampling frequency was used to collect waste samples and accredited laboratory of PCRWR-Islamabad was used for analysis purpose. Statistical tools were used to test the spatial and temporal variations of wastewater in the study area. Average daily waste flows from each catchment of Bharakaho generated from different sources were estimated by using different empirical relationships as described below.

3.2.1. Domestic waste flow estimations

The relationship in Eq. 1 was used for the estimation of domestic waste flow from the study area.

$$Q_{DW} = P_t \times q \tag{1}$$

Where

Q_{DW} = Domestic Sewage Flow (lit/day)

P_t =Total Population of the Area

Q = Average daily per capita water use (lit/day)

75% of water supply was assumed to be returning as sewage as suggested by Vesilind [32].

3.2.2. *Industrial waste flow estimations*

Industrial waste estimation was inclusive of i) waste generated per employee (17.5 GD^{-1}) and ii) waste generated per square foot (0.18 GD^{-1}) as recommended by the University of Minnesota [33].

Total industrial waste generation per day from industrial campus = Waste generated by employees + Waste generated from total area:

$$W_{TI} = W_E + W_{TA} \tag{2}$$

Where

W_{TI} = Total industrial waste generation per day from industrial campus (unit)

W_E = Waste generated by employees (unit)

W_{TA} = Waste generated from total area (unit)

3.2.3. *Commercial waste flow estimation*

Commercial waste flow estimation was dependent upon type of commercial activity. For example, from hospitals, the waste flow was estimated using following formula after [33]:

Total waste generated from hospitals = Waste generated by patients + Waste generated by practitioners + Waste generated from total area

Waste generate per patient = 3 GD^{-1}

The number of patients was known by visiting hospitals.

Waste generated by practitioners = Waste generated per practitioner x Number of practitioners

Waste generated per practitioner = 275 GD^{-1} [33]

The number of practitioners is known by visiting hospitals.

Waste generated from total area = Waste generated per square foot x Area of commercial activity

Waste generated per square foot = 1.1 GD^{-1} [33]

Area of commercial activity is known through visiting the commercial activity. Similarly, wastes are generated from all commercial activities. The waste is often extensive in few drains depending upon the urban development in the area (Figure 3).

Average daily waste loading (organic and inorganic) was estimated to quantify the amount of waste being generated from the study area, so that proper wastewater treatment technology can be recommended for the area. Wastewater treatment options were analyzed and a criterion was developed that included costs (capital, operating, and maintenance), technology, manpower, climatic conditions, and community interactions.

Figure 3. Urbanization in Bharakahu (left) and wastewater from Hathala drain entering Korang River (right).

4. Results and discussion

4.1. Water and wastewater discharge pattern

Through questionnaire-based surveys conducted in the pilot area of Bharakaho, information related to water supply, drainage mode, water consumption patterns, and existing wastewater systems were gathered. In majority of the area (about 60%), groundwater was the major source of water supply followed by surface-tapped water from Simly dam (26.3%) for domestic needs (Figure 4). While a few were using both surface and sub-surface water supplies for domestic purpose (6.3%). When inquired about average water consumption, majority was using 100–200 liters of water per person per day (56.6%), while significant proportion of inhabitants (21.7%) also reported less than 100 liter/day/person water consumption. Some also were lucky enough to have excess of more than 300 liters. This trend overall suggested that water scarcity was not an issue and people were getting quite handsome amount of water supply. When inquired about waste domestic drainage mode, more than half of the population was using buried pipe lines (62.3%), while 32.6% people were using open drains to discharge wastewaters out of homes. Further, it was revealed that almost entire population was using septic tanks as a means of preliminary treatment. Further analysis indicated that combined sewer system was prevailing in the area (95%) and awareness regarding untreated wastewater discharges into freshwater streams was very high (77%). While interesting to note was the fact that majority (about 71%) were willing to pay for wastewater treatment facilities if provided.

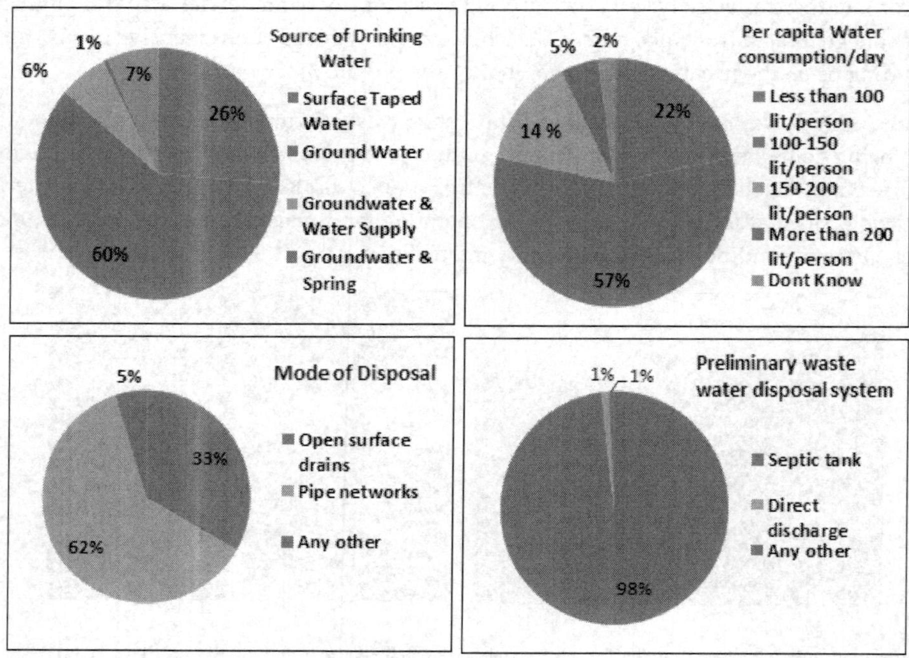

Figure 4. Water consumption and discharge patterns in the studied catchments of Bharakaho.

4.2. Wastewater production potential

Wastewater was not only being produced from domestic sources but there were significant industrial (construction, marble industry, and commercial activities from hospitals, markets, school, etc.) sources within the study area that were also consuming fresh water supplies and discharging wastewaters into freshwater streams. The estimated wastewater production potential per day from all these sources has been summarized in Table 2. Out of 6.354 million gallons per day (MGD) waste flow discharging from the study area, 97.4% was being added from the domestic sources, while 2.6% was contributed by industrial and commercial sources. It increases significantly at each step as river water approaches towards the lake. The Kiani road catchment was producing most non-domestic sewage flows (0.63 MGD), while it was also thickly populated area and producing highest levels of domestic sewage (2.377 MGD) followed by Hathala catchment (1.783 MGD). The wastewater sampling results are shown in Table 3. The values were averaged over 14 months of study period.

The permissible limits of various wastewater parameters for Pakistan are depicted in Table 4. These limits of the parameters are National Environment Quality (NEQ) standards for wastewater parameters to be discharged in water or on land [34]. By comparing the actual parametric values with the NEQ standards, it is obvious that Shahdara (before bridge), Col. Amanullah road, and Hathala catcments were discharging BOD more than permissible levels, while COD was only exceeding for Hathala catchment and all other parameters were within

Location	Domestic Waste (MGD)	Industrial and Commercial Waste (MGD)	Total Waste (MGD)
Shahdara catchment (Before bridge)	0.470	0.005	0.475
Col. Amanullah Road catchment	0.966	0.014	0.980
Hathala catchment	1.783	0.041	1.824
Kiani road catchment	2.377	0.063	2.440
Shahdara catchment (After bridge)	0.594	0.041	0.635
Total	**6.190**	**0.164**	6.354

Table 2. Wastewater production potential from different sources in the study area.

permissible limits. However, these standards were for wastewaters, while the storage of Rawal dam was being used for drinking purposes, which required zero BOD levels, and for that reason this was unfit water.

Once waste flow rates and concentrations of pollutants were estimated, the actual water loading rates in freshwater streams were calculated (Table 5). A net BOD of 2,296 kg day^{-1} and COD of 3,875 kg day^{-1} were being discharged from the area, which was considered very high. Nitrate was being discharge at highest levels (141 kg day^{-1}), followed by phosphorus related pollutants (53 kg day^{-1}), while TDS were being added at the rate of 19,653 kg day^{-1} through the study area via domestic, commercial, and industrials means (Table 5). The data indicated that spatial variability was also prevalent in the study area. For example, Col. Amanullah catchment was highly polluted due to higher values of BOD and COD (dense population and closed nature of catchment has reduced surface runoff from outside non-urbanized area), while Shahdara catchment (after bridge) was least problematic due to having large open rangeland/vegetation cover in the catchment. According to Kahlown et al. [35] increase in population has significant effect on the water quantity and quality as the increase in population is a cause of increase in contaminants and some other wastewater parameters.

Location	BOD (ppm)	COD (ppm)	pH	EC (μScm^{-1})	TP (ppm)	Nitrate (ppm)	Nitrite (ppm)	TDS (ppm)
Shahdara catchment (before bridge)	52	16	7.29	359	0.45	4	0.89	209
Col. Amanullah road catchment	74	66	7.06	525	1.35	2	0.045	315
Hathala catchment	112	143	6.93	631	0.68	3	0.015	378
Kiani road catchment	48	36	6.95	617	1.15	8.5	1.875	371
Shahdara catchment (after bridge)	3	3	7.09	400	0.33	4.5	1.675	239

Table 3. Average pollution loadings from the study area.

Wastewater Parameters	Permissible Limits
Biochemical Oxygen Demand	50 mgl^{-1}
Chemical Oxygen Demand	100 mgl^{-1}
Electrical Conductivity	1,000 μScm^{-1}
pH	6.0–8.0
Nitrate	20 mgl^{-1}
Nitrite	2 mgl^{-1}
Total Phosphates	10 mgl^{-1}
Total Dissolved Solids	1,200 mgl^{-1}

Table 4. Permissible limits of wastewater parameters as per NEQ standards.

Locations	BOD ($kg.day^{-1}$)	COD ($kg.day^{-1}$)	TP ($kg.day^{-1}$)	Nitrate ($kg.day^{-1}$)	Nitrite ($kg.day^{-1}$)	TDS ($kg.day^{-1}$)
Shahdara catchment (before bridge)	98	158	3	6	0.17	963
Col. Amanullah Road catchment	680	1117	11	30	0.29	2,961
Hathala catchment	547	1052	16	50	7.25	6,649
Kiani road catchment	940	1513	22	49	1.29	8,225
Shahdara catchment (after bridge)	31	35	1	6	1.73	855
Total	2,296	3,875	53	141	4.73	19,653

Table 5. Estimated loading rates of BOD, COD, TP, Nitrate, Nitrite, and TDS.

Pollution parameters such as BOD, COD, phosphates, and TDS were being discharged in large quantities (940, 1513, 22, and 8,225 $kg.day^{-1}$ respectively) from Kiani road catchment, while nitrate (50 $kg.day^{-1}$) and nitrite were being added (7.25 $kg.day^{-1}$) from the Hathala catchment. The values of all these parameters were lowest in the Shahdara catchment.

4.3. Impact of seasonality on wastewater

Careful analysis of rainfall and pollution loading rates (BOD & COD) were following reverse interaction throughout the study period. The rainfall recorded in Satrameel field station located in the study area has shown an increasing pattern during monsoon months from July to September (Figure 5). The lowest values for both BOD and COD were observed during the wet period of monsoon when increased surface runoff diluted the pollutions (least in September), while higher concentrations were recorded during dry periods (highest during March) when after monsoon even base flows started decreasing (Figure 6). Although higher rainfalls during June to September 2011 had caused a lot of surface runoff that resulted in lowering of the pollution levels, nitrate and nitrite increased in amount due to the washing out of human, poultry, and animal wastes.

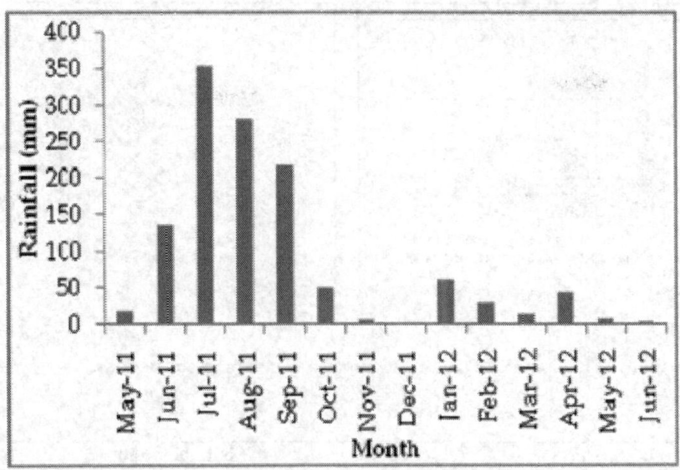

Figure 5. Monthly rainfall pattern during study period.

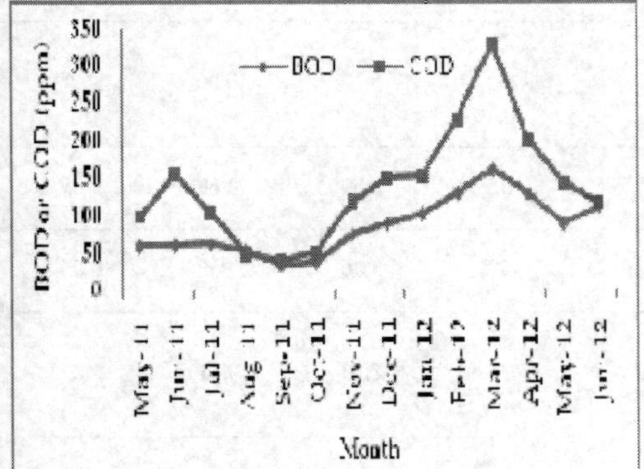

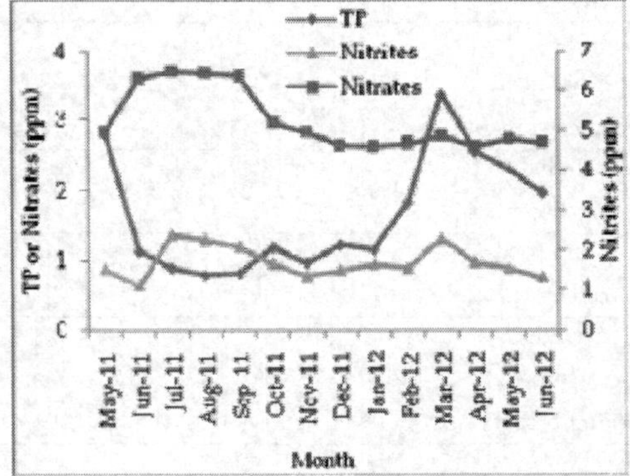

Figure 6. Temporal behavior of BOD, COD, TP, nitrite, and nitrate in the study area.

To trace seasonality, further analysis was carried out and outcome of BOD and COD is shown in Figure 7. Seasonality has indicated a clear impact on both selected parameters during study during wet period of Monsoon increased freshwater surface runoff in natural streams caused significant dilution and consequently reduced concentrations of BOD and COD were observed for all locations. While in contrary, during non-monsoon periods, the corresponding values were much higher for respective catchments. Similarly, the other pollution parameters were also analyzed in context of seasonality (Table 6). A close insight of both tables revealed that generally pH increased with rains (monsoon) mainly because runoff waters brought large quantities of wastes, while phosphorus and nitrate increased at some catchments during monsoon due to increased transport of nutrients with runoff water. EC, on the other hand, decreased during the monsoon period due to dilution impact.

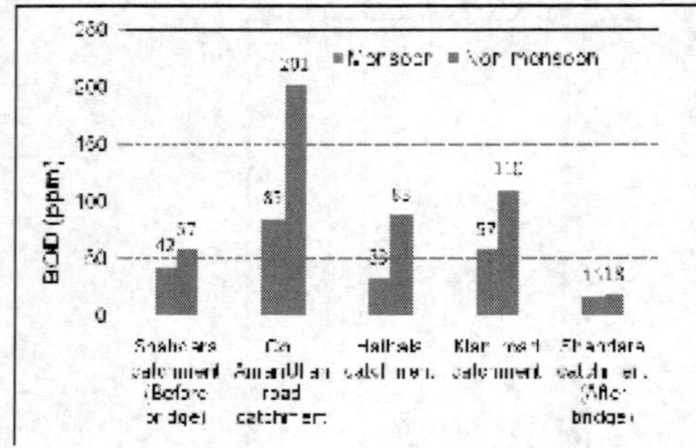

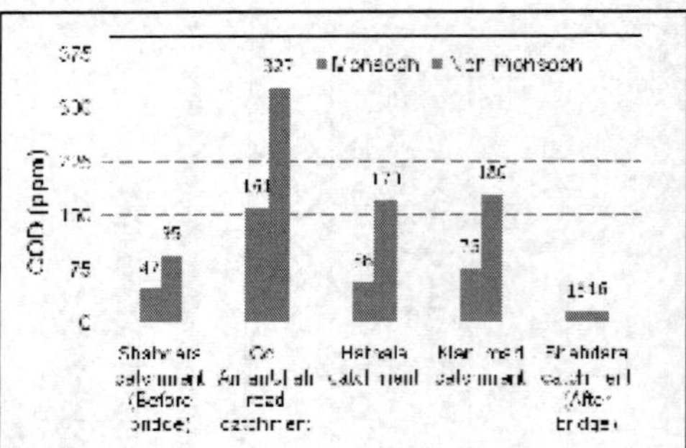

Figure 7. Spatial variability of BOD and COD during monsoon and non-monsoon periods.

Location	EC (μScm⁻¹)		TP (ppm)		Nitrate (ppm)		Nitrite (ppm)	
	Monsoon	Non-monsoon	Monsoon	Non-monsoon	Monsoon	Non-monsoon	Monsoon	Non-monsoon
Shahdara catchment (before bridge)	762	926	1.1	1.61	3.5	3.6	0.03	0.11
Col. Amanullah road catchment	1,303	1,343	3.25	2.6	5	8.73	0.06	0.08
Hathala catchment	1,410	1,793	1.7	2.25	10.5	6.59	4.69	0.39
Kiani road catchment	1,226	1,468	2.71	2.39	4.75	5.46	0.6	0.05
Shahdara catchment (after bridge)	580	595	0.75	0.25	4.1	2.33	0.11	0.82

Table 6. Impact of seasonality on EC, TP, nitrate, and nitrite.

Statistical analysis of One-Way ANOVA technique was employed to determine significance of seasonality for various parameters. BOD indicated higher values of P (>0.1) and coefficient of variance (CV=105.3) for temporal scale and lowest values (P<0.05 & CV=73.2) for spatial scale exhibiting more variability with space than time. Similar trends were also found for COD and other major parameters. However, pH varied both spatially and temporally while Nitrites were neither temporally not spatially varied.

4.4. Water quality response to land use change scenarios

There have been occurred extensive land use changes in the watershed during the last two decades resulting from deforestation and high growth in the urbanization [14]. The responses of sediment yield and water quality parameters, e.g., soluble N and P and nitrate contribution to reach, to changes in the land use were studied. The land use/land cover extent estimated in Rawal watershed during base year and under various scenarios is shown in Table 7. These scenarios are intended to be prospective and informative rather than projective or prescriptive of the future [36]. Scenario-1 is related to deforestation case in which all the scrub forest is assumed to be converted into rangeland (the rangeland increases to 75.5%). The natural forests in the country have been subjected to deforestation for growing agricultural crops, grazing domestic animals, and obtaining fuel wood and timber for the last many years [37]. The extensive grazing and cutting of wood have deformed the plants into bushes [21]. Scenario-2 represents the case of increase in agricultural development and growth in cropping activities. All the rangeland of base year is assumed to be converted into agriculture land (the agriculture land increases to 44.1%) in this scenario. Non-point sources particularly from agriculture are generally the major causes of nutrient pollution [38]. Scenario-3 is related to case of growth in urbanization under which all the rangeland of base year is assumed to be converted into built-up land, i.e., it increases to 45.6% in the watershed.

In scenario-1, the surface runoff has shown an average increase of about 0.9%, while sediment yield increases by about 26% from that of the base year 2010. The organic N and P exhibit more or less same positive change of about 23% in this scenario. The contribution of nitrate to stream flow increases slightly (about 1%) due to degradation of the scrub forest.

In scenario-2, the surface runoff indicates an increase ranging between 0–10.1%, while sediment yield increases on an average by about 21%. The organic N exhibits an average decrease of about 1.9%, while organic P increases by 3.6% due to growth in the agriculture developments. The contribution of nitrate to stream flow increases on an average by 2.4% in the watershed (Table 8).

Land use	Base year	Senario-1	Senario-2	Senario-3
Conifer	2.3	2.3	2.3	2.3
Scrub	38.8	0	38.8	38.8
Agriculture	7.4	7.4	44.1	7.4
Rangeland	36.7	75.5	0	0
Soil/Rocks	4.8	4.8	4.8	4.8
Settlement	8.9	8.9	8.9	45.6
Water	1.1	1.1	1.1	1.1
Total	100	100	100	100

Table 7. Percentage extent of land use/land cover in base year and under various land use scenarios.

The surface runoff has shown an increase of about 3.1% in scenario-3, likely due to expansion in the imperviousness. The upper sub-basins of the watershed indicate an increase in the surface runoff under scenarios -2 and -3, the runoff being higher in the later scenario due to urban development (Figure 8). There is a minor decrease in the sediment yield (about 4.1%)

in scenario-3 that may be attributed to the decrease in the rangeland, e.g., grass/shrubs that is replaced by the built-up land. The increase in the sediment yield is prominent under scenarios -1 and -2 (the cases of deforestation and agriculture development) particularly in the upper sub-basins (Figure 9). In scenario-3, the contribution of nitrate to stream flows shows a slight increase from that of scenario-2, overall presenting an identical picture of distribution in different sub-basins of the watershed under these two scenarios (Figure 10).

Scenario	Parameter	SURQ	SYLD	NSURQ	ORGN	ORGP
	Max	4.311	165.759	5.247	129.144	128.692
1	Min	0.000	0.000	0.000	0.000	0.000
	Average	0.869	26.337	1.045	23.422	23.167
	Max	10.086	112.104	11.485	0.000	20.659
2	Min	0.000	0.000	0.000	-16.218	0.000
	Average	2.128	20.994	2.407	-1.919	3.609
	Max	14.375	0.000	12.509	0.000	0.000
3	Min	0.000	-30.002	0.000	-27.725	-27.057
	Average	3.111	-4.120	2.589	-4.322	-4.148

SURQ = Surface runoff; SYLD = Sediment yield; ORGN = Organic-Nitrogen; ORGP = Organic Phosphorus; and NSURQ = Nitrate contribution to stream flow

Table 8. Percentage change in the water quality parameters from that of base year.

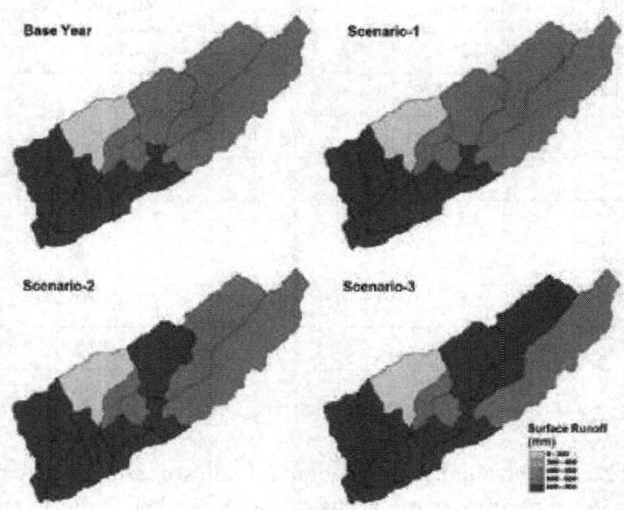

Figure 8. Surface runoff under various scenarios of land use change.

The changes in various parameters in the watershed under three scenarios are shown in Figures 8–12. The organic N and P indicate an average decrease of about 4% likely due to growth in urbanization in scenario-3. The results of organic N are similar to scenario-2 but differ from scenario-1 that indicates a positive change in the upper sub-basins of the watershed (Figure 11). The changes in organic P are diverse in various sub-basins under all three scenarios, being less significant under scenario-3 due to high growth in the urbanization (Figure 12).

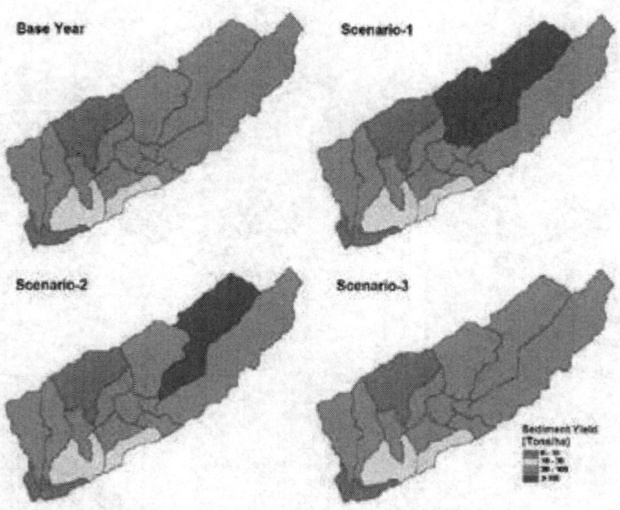

Figure 9. Sediment yield under various scenarios of land use change.

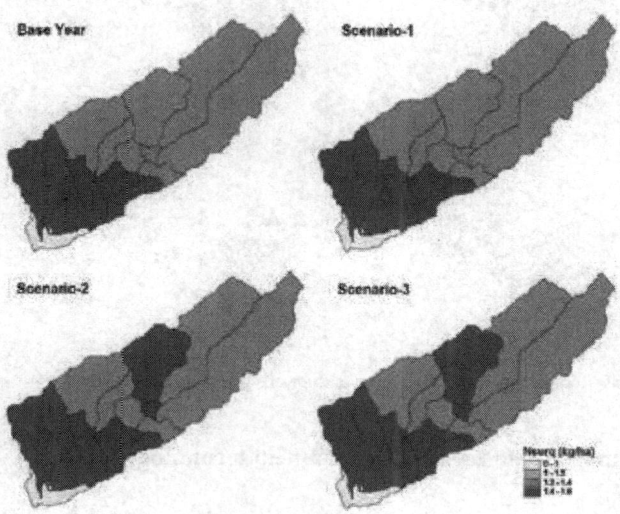

Figure 10. Contribution of nitrate to reach in different sub-basins of the watershed.

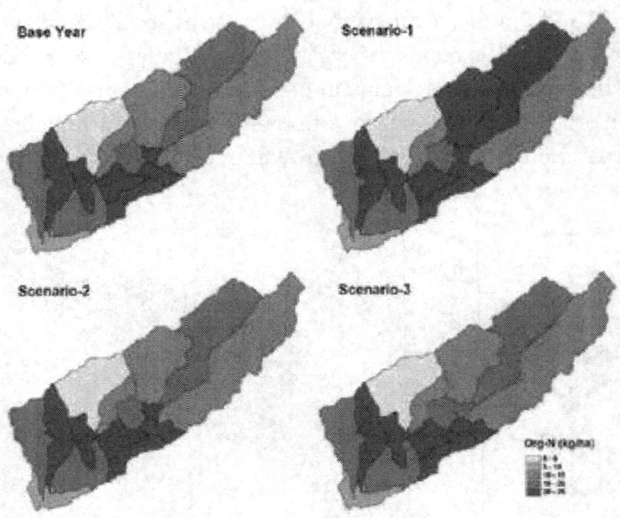

Figure 11. Organic nitrate generated under various scenarios.

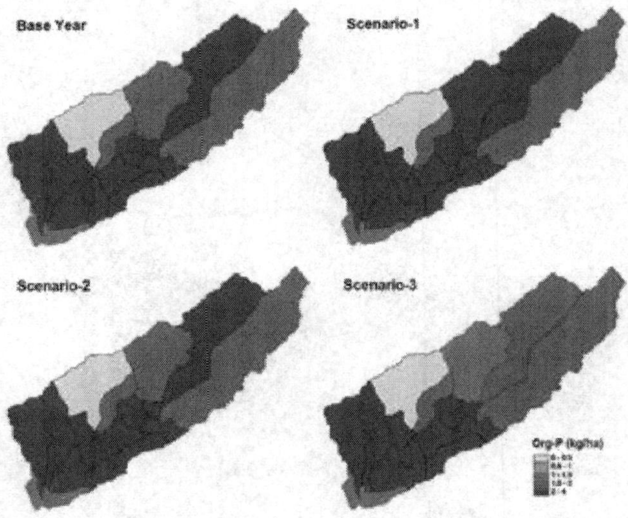

Figure 12. Organic phosphorous generated in different sub-basins under three scenarios.

4.5. Selection of appropriate wastewater treatment technology

Properly planned use of municipal wastewater alleviates surface water pollution problems and not only conserves valuable water resources but also takes advantage of the nutrients contained in sewage to grow crops. Based upon the above analysis, it was evident that untreated wastewater discharge towards Rawal dam was a serious issue due to drinking nature of the storage and remedies were essential. A scientifically designed septic tank with multiple compartments was considered to be the most feasible wastewater management solution due to typical geographical and socio-economic conditions of the study area. Although over 98% area, already these were prevailing but these were just dug holes without considering any scientific justifications. If a properly planned septic tank could be developed and replaced in entire area, it can reduce the wastewater contamination issue at Rawal dam very significantly. As septic tanks can be 60%–70% efficient in reducing wastewater loads, dilution through freshwaters from Rawal watershed area will help in mitigating remaining impurities. To address these issues, conservation practices, such as conservation tillage, filter strips, land retirement, and nutrient management can be adopted [25]. Other policy level planning should be to (a) encourage reduction, recycling, and reuse of municipal and industrial solid and liquid wastes; (b) develop and enforce rules and regulations for proper management of municipal, industrial, hazardous, and hospital wastes; (c) develop environmental risk assessment guidelines for existing industries as well as new development interventions; and (d) adopt cleaner technology, implement pollution control measures, and compliance with the environmental standards.

5. Conclusions

The issues related to water quality and quantity management involves numerous technical, socio-economic and environmental factors and coupled with complex spatial variability become highly complicated. The results of the integrated field based investigations and modeling approach exhibited variable responses of surface runoff and water quality to changing conditions of urbanization and deforestation. The concentration of the pollution parameters was found higher from the densely populated catchments containing extensive industrial and commercial activities. Temporal wastewater analysis indicated a significant impact of seasonality on the contaminants' population levels. The wet season of monsoon (July–September) has lowered almost all key parameters of pollution as compared to the dry season, except nitrate and nitrite, which indicated increasing trends during the wet period because runoff also taken animal, human, and poultry wastes dumped in or near the drainage network. Statistically, the variability was significant on the temporal scale and non-significant on the spatial scale—an indication that wastewater contamination is affected by the seasons. If planned properly, the municipal wastewater can be used not only for conserving valuable water resource, but also for taking advantage of the nutrients contained in sewage to grow crops. It is desirable to develop mathematical techniques to aid decision makers in formulating cost-effective and environment-friendly plans and policies for wastewater management. The modeling efforts using SWAT biophysical water quality model would enhance the capability of decision makers in exploring comprehensive and ambitious plans for managing water systems. The adopted approach would provide an effective decision support tool for evaluating better management options to reduce negative impacts of wastewater and contaminants for sustainable agro-environment in future.

References

[1] Li Y, Huang G, Huang Y, Qin X. Modeling of water quality, quantity, and sustainability. Journal of Applied Mathematics Vol. 2014 (2014), Article ID 135905. 3p. http://dx.doi.org/10.1155/2014/135905.

[2] Beck MB, Reda A. Identification and application of a dynamic model for operational management of water quality. Water Science & Technology. 1994; 30(2): 31-41.

[3] Wrede S, Seibert J, Uhlenbrook S. Distributed conceptual modelling in a Swedish lowland catchment: a multi-criteria model assessment. Hydrology Research. 2012; (published on-line) DOI:10.2166/nh.2012.056.

[4] Daloğlu I, Nassauer JI, Riolo R, Scavia D. An integrated social and ecological modeling framework—impacts of agricultural conservation practices on water quality. Ecology and Society. 2014; 19(3): 12. http://dx.doi.org/10.5751/ES-06597-190312.

[5] Zhang W, Jia Q, Chen X. Numerical Simulation of Flow and Suspended Sediment Transport in the Distributary Channel Networks. Journal of Applied Mathematics Vol. 2014 (2014), Article ID 948731, 9p. http://dx.doi.org/10.1155/2014/948731

[6] Veldkamp A, Verburg PH. Modelling land use change and environmental impact. Journal of Environmental Management. 2004; 72(1–2): 1-3.

[7] Levin S, Xepapadeas T, Crépin AS, Norberg J, de Zeeuw A, Folke C, Hughes T, Arrow K, Barrett S, Daily G, Ehrlich P, Kautsky N, Maler KG, Polasky S, Troell M, Vincent JR, Walker B. Social–ecological systems as complex adaptive systems: modeling

and policy implications. Environment and Development Economics. 2012; 18(2): 111-132.

[8] Nesmerak I, Blazkova SD. Analysis of the time series of waste water quality at the inflow of the wastewater treatment plant and transfer functions. Journal of Hydrology and Hydromechanics. 2014; Vol. 62(1): 55-59. DOI: 10.2478/johh-2014-0009.

[9] Malik S. Rawal dam floating on Garbage. Published in Daily Times on December 28, 2005.

[10] Sobia. Determination of Pesticide Residues in Rawal and Simly Lake, A thesis submitted to fulfill partial requirement of degree of Masters in Environment in Department of Environmental Sciences, Fatima Jinnah Women University, Rawalpindi, Pakistan; 2007.

[11] Ahmad I, Anwer MD, Ahmad Z. Reservoir Water Quality Monitoring in Rawal Lake using Geoinformatics. Pakistan Journal of Science. 2012; 64(1).

[12] Ashraf A, Ahmad Z. Integration of Groundwater Flow Modeling and GIS, in Water Resources Management and Modeling, Dr. Purna Nayak (Ed.), ISBN: 978-953-51-0246-5. InTech; 2012. p. 239-262.

[13] Yasin A. Rawal Lake Water Purification. An editorial published in Daily Times on August 20, 2009.

[14] Ashraf A, Naz R, Syial AW, Ahmad B, Yasin M, Saleem M. Assessment of landuse change and its impact on watershed hydrology using remote sensing and SWAT modeling techniques. Int. Jour of Agri. Sc. and Tech. 2014; 2(2): 61-68. DOI: 10.14355/ijast.2014.0302.02.

[15] Arnold JG, Srinivasan R, Muttiah RS, Williams JR. Large area hydrologic modeling and assessment - Part 1: Model development. Journal of the American Water Resources Association. 1998; 34(1): 73-89.

[16] Arabi M, Frankenberger JR, Enge BA, Arnold JG. Representation of agricultural conservation practices with SWAT. Hydrological Processes. 2008; 22(16): 3042-3055.

[17] Neitsch SL, Arnold JG, Kiniry JR, Williams JR. Soil and water assessment tool (SWAT) theoretical documentation. Texas A&M University, College Station, Texas, USA. 2011.

[18] Ribaudo MO, Smith, MM. Water quality: impacts of agriculture. In AH-722. Agricultural Resources and Environmental Indicators. Economic Research Service, USDA, Washington, D.C., USA. 2000.

[19] Boyer EW, Goodale CL, Jaworsk NA, Howarth RW. Anthropogenic Nitrogen sources and Relationships to Riverine Nitrogen export in the Northeastern USA. Biogeochemistry. 2002; 57(1): 137-169.

[20] Galloway JN, Dentener FJ, Capone DG, Boyer EW, Howarth RW, Seitzinger SP, Asner GP, Cleveland CC, Green PA, Holland EA, Karl DM, Michaels AF, Porter JH,

Townsend AR, Vöosmarty C J. Nitrogen cycles: past, present, and future. Biogeochemistry. 2004; 70(2): 153-226.

[21] Shafiq M, Ahmed S, Nasir A, Ikram MZ, Aslam M, Khan M. Surface runoff from degraded scrub forest watershed under high rainfall zone. Journal of Engineering and Applied Sciences. 1997; 16(1): 7-12.

[22] Aftab N. Haphazard colonies polluting Rawal Lake, Daily Times Monday, March 01, 2010. http://www.dailytimes.com.pk/default.asp...009_pg11_1.

[23] Ghumman AR. Assessment of water quality of Rawal Lake by long-time monitoring. Environ Monitoring and Assessment. 2010; 180: 115-126.

[24] PEPA (Pakistan Environmental Protection Agency). Report on Rawal Lake Catchment Area Monitoring Operation. Ministry of Environment, Islamabad, Pakistan. 2004: p.19.

[25] Dolan DM, Chapra SC. Great Lakes total phosphorus revisited: 1. Loading analysis and update (1994–2008). Journal of Great Lakes Research. 2012; 38(4): 730-740.

[26] Saeed MA, Ashraf A, Ahmad B, Shahid M. Monitoring deforestation and urbanization growth in Rawal Watershed area using Remote Sensing and GIS techniques. Jour of COMSATS- Science Vision. 2011; Vol. 16&17: 93-104.

[27] IUCN. Rapid environmental appraisal of developments in and around Murree Hills, IUCN Pakistan. 2005.

[28] Hawley N, Johengen TH, Rao YR, Ruberg SA, Beletsky D, Ludsin SA, Brandt SB. Lake Erie hypoxia prompts Canada–U.S. study. Eos, Transactions of the American Geophysical Union. 2006; 87(32): 313-319.

[29] Zhou Y, Obenour DR, Scavia D, Johengen TH, Michalak AM. Spatial and temporal trends in Lake Erie hypoxia, 1987–2007. Environmental Science and Technology. 2013; 47(2): 899-905. http://dx.doi.org/10.1021/es303401b.

[30] Scavia D, Allan JD, Arend KK, Bartell S, Beletsky D, Bosch NS, Brandt SB, Briland RD, Daloğlu I, DePinto JV, Dolan DM, Evans MA, Farmer TM, Goto D, Han H, Höök TO, Knight R, Ludsin SA, Mason D, Michalak AM, Richards RP, Roberts JJ, Rucinski DK, Rutherford E, Schwab DJ, Sesterhenn T, Zhang H, Zhou Y. Assessing and addressing the re-eutrophication of Lake Erie: Central basin Hypoxia. Journal of Great Lakes Research. 2014; 40(2): 226-246.

[31] Carpenter SR. Phosphorus control is critical to mitigating eutrophication. Proceedings of the National Academy of Sciences of the United States of America. 2008; 105(32): 11039-11040. DOI:10.1073/pnas.0806112105.

[32] Vesilind AP. Wastewater Treatment Plan Design. A book published in Water Environment Federation. 2003.

[33] University of Minnesota publication. Onsite Sewage Treatment Plant - Wastewater Sources and Flows. Water Resource Center. 2011.

[34] The Government Gazette of National Environmental Quality Standards (NEQS). Government of Pakistan. 2000.

[35] Kahlown MA, Ashraf M, Hussain M, Abdul Salam H, Bhatti AZ. Impact assessment of sewerage and industrial effluents on water resources, soils, crops and human health in Faisalabad. A report published by Pakistan Council of Research in Water Resources (PCRWR), Islamabad, Pakistan. 2006: p.110.

[36] Nassauer JI, Corry RC. Using Normative scenarios in Landscape Ecology. Landscape Ecology.2004; 19 (4): 343-356.

[37] Siddiqui KM, Mohammad I, Ayaz M. Forest ecosystem climate change impact assessment and adaptation strategies for Pakistan. Climate Research. 1999; 12: 195-203.

[38] Forster DL. Public policies and private decisions: Their impacts on Lake Erie water quality and farm economy. Journal of Soil and Water Conservation. 2000; 55(3): 309-322.

Index